Science as Child's Play in Seventeenth-Century England

Elizabeth L. Swann

Science as Child's Play in Seventeenth-Century England

Innocence, Experience, Experiment

Elizabeth L. Swann
Department of English Studies
Durham University
Durham, UK

ISBN 978-3-031-75848-5   ISBN 978-3-031-75849-2   (eBook)
https://doi.org/10.1007/978-3-031-75849-2

For
Edith, Kit, and Louisa.

ACKNOWLEDGEMENTS

This short book has its own infancy in an essay commissioned for the volume *Experiential & Experimental Knowledge on the Early Modern English Stage*, edited by Pavneet Aulakh and James Kearney (forthcoming with Edinburgh University Press). I am grateful to the editors for the impetus to write the piece, which swiftly expanded beyond the parameters of an essay. I am also grateful to the editors, and to Edinburgh University Press, for permission to include some of the material from the essay in this much-expanded version in advance of the edited volume's publication.

For productive conversations about some of the ideas in this book at a germinal stage, I am grateful to Ofer Gal, Mariam Jacobson, Yaakov Mascetti, Emilie Murphy, and Stephanie Shirilan. In May 2022, I delivered a portion of an early version of the manuscript at an online seminar hosted by Concordia University's *Centre for Sensory Studies*; I am grateful to David Howes for this invitation and to the audience for fruitful discussion. Pavneet Aulakh and the late Bruce Smith provided helpful feedback on the original essay that led to this book. The anonymous reviewer of the manuscript at Palgrave Macmillan also offered both valuable encouragement and constructive critique. Paula Findlen read the manuscript at a late stage. I am extremely grateful for her generous and incisive suggestions; those which I haven't been able to incorporate here will certainly inform future projects.

Thanks, too, are due to family and friends. My father, Gregory Swann, is unstinting in his love and support, which takes more forms than I can list here. My mother, Louise Swann, made my return to work possible after the birth of my third child in 2023, leaving her own home, partner, and job and moving hundreds of miles to Durham to provide full-time domestic care for a year. I am so grateful to them both. Thanks also to my excellent in-laws, Elizabeth and Ian McAuley, who have helped in many ways.

I am grateful to my husband, Tomás McAuley, for his continued love and support. My own chaotic, gorgeous little scientists—Edith, Kit, and Louisa McAuley-Swann—are a continuous source of joy, as well as pine-cones and pebbles. This book is dedicated to them, with love.

Note on the Text

In my citations of early modern sources, I modernize i/j, u/v, y/i, long s and double v, and silently expand contractions, but retain original punctuation, spellings, and italicization. I have also capitalized or uncapitalized silently the initial letter in a quotation when this is required by the sentence structure. In the notes and bibliography, I have shortened many longer titles.

CONTENTS

LIST OF FIGURES

CHAPTER 1

Introduction: "Things only fit for Children"

Abstract This introduction outlines the main themes and scope of the work, situating it in relation to relevant scholarship by figures including Paula Findlen and Steven Shapin. It opens with a discussion of seventeenth- and eighteenth-century satirical representations of the experimental science associated with the Royal Society of London, including Margaret Cavendish's *Observations* and Thomas Shadwell's *The Virtuoso*. In representing that science as childish, I suggest, such satires capitalize on a connection that predecessors, members, and supporters of the Society frequently made themselves. In the early modern period, childhood was understood variously as state of corruption and error, and innocence and insight. Both these strands of thought inflect the work of figures including Francis Bacon, Robert Boyle, and Robert Hooke. On the one hand, experimentalism is associated with manly maturity. On the other, it is valorized as an innocent and epistemically productive form of childish play. The introduction concludes with a discussion of Samuel Pepys' description of Cavendish's attendance at a meeting of the Royal Society, where the presence of children is noted by Pepys but rarely discussed by modern scholars.

Keywords Science • Childhood • Natural philosophy • Experiment • Satire • Original sin • Innocence • Royal Society • Francis Bacon • Margaret Cavendish • Samuel Pepys • Thomas Shadwell • William King

© The Author(s), under exclusive license to Springer Nature Switzerland AG 2024
E. L. Swann, *Science as Child's Play in Seventeenth-Century England*, https://doi.org/10.1007/978-3-031-75849-2_1

1

In the chapter on the microscope included in her *Observations upon Experimental Philosophy* of 1666, Margaret Cavendish mocks "Experimental Philosophers" by comparing them to "boys that play with watery bubbles or fling dust into each other's eyes, or make a hobbyhorse of snow". The marginal notes offer more specificity: the "bubbles" are glossed as "glass-tubes" whilst "dust" represents "atoms", and the hobby horses are "exterior figures". Disparaging the work of the recently formed Royal Society of London for Improving Natural Knowledge as puerile and fruitless play, Cavendish concludes that such "are worthy of reproof rather than praise, for wasting their time with useless sports".[1]

Ten years later, in Thomas Shadwell's satire on the Royal Society, *The Virtuoso* (1676), the vain and frivolous Lady Gimcrack complains of incompatibility with her husband Sir Nicholas—often taken to represent Robert Hooke—on the grounds of a disparity in age and temperament.[2] Although Sir Nicholas is "a fine solitary Philosophical person", she preens, "my nature more affects the vigorous gaiety and jollity of Youth, than the fruitless speculations of Age".[3] Lady Gimcrack's depiction of her husband's experimental obsessions as evidence of his decrepitude, however, is belied by the infantile tenor of his activities, from playing with his "tame Spider" to attempting to fly.[4] In the following scene, we encounter the ridiculous Sir Nicholas learning to swim using a somewhat unconventional experimental method: "He has a Frog in a Bowl of Water, ty'd with a pack-thred by the loins; which pack-thred Sir Nicholas holds in his teeth, lying upon his belly on a Table; and as the Frog strikes, he strikes; and his swiming-Master stands by, to tell him when he does well or ill".[5] Challenged by his nieces' suitors about his refusal "to practise in the Water", he dismisses their suggestion that swimming is pointless on dry land: "I seldom

---

[1] Cavendish, *Observations*, D2r (11). On Cavendish's suspicion of games such as tennis, chess, cards and dice, and on her endorsement of the "imaginative play" of world-building, see Nelson and Alker, "Virtual Reality, Role Play and World Building", 117–137. For an account of Cavendish's critique of the Royal Society which emphasizes its gendered aspects, see Dear, "A Philosophical Duchess".

[2] Hooke's response to Shadwell's work was, famously, furious: in a diary entry written on 27 June 1675, he called it an "atheistical wicked play", and the following year, on 2 June 1676, he fulminated about a performance he attended: "Damned Doggs. *Vindica me Deus.* People almost pointed". Hooke, *The Diary*, 166, 235.

[3] Shadwell, *The Virtuoso*, D3v (24).

[4] Shadwell, *The Virtuoso*, G1r (43).

[5] Shadwell, *The Virtuoso*, D4r (25).

bring any thing to use", he proclaims, "'tis not my way".[6] Thrashing about on a table, a novice play-acting under the watchful eye of a teacher, Sir Nicholas—whose very name, Gimcrack, aligns him with the "Knacks", or toys, that fill "his Laboratory"—illustrates the indignities of senility and the silliness of youthful play simultaneously.[7] In a play predicated on generational conflict, then, Lady and Sir Nicolas Gimcrack stand out for their refusal to act their age: the former, conventionally enough, by courting the attention of younger men; the latter, more strikingly, by indulging in a childish experimentalism unfit for the decorum of his advanced years.

At the beginning of the next century, the joke still hadn't got old. In the poet and satirist William King's influential dialogue *The Transactioneer* (1700), a figure representing the physician and naturalist Sir Hans Sloane, Secretary to the Royal Society and editor of the *Philosophical Transactions*, attempts to impress a sceptical "Gentleman" by boasting that "the Remarkable Passages taken Notice of" in his "Philosophical News Papers" include "an Account of *Glow-Worms Volant* [capable of flight], and *Butterflies Eggs that were testaceous* [having a shell], *and near as big as Wren's, most gloriously bestudded with Gold and Silver*", which "*Hatch in the Windows, and are a sport for Children*".[8] The Gentleman is unconvinced: "Pray", he asks sardonically, "how came your Correspondent to take notice of Things only fit for Children; What? did he think your Genius lay the same way as Childrens do?"[9] Puncturing the pretension of the correspondent's grandiloquent jargon, the Gentleman ridicules him and his peers in now-familiar terms, depicting their entomological fixations as risibly juvenile.

In King's satire, however, the Transactioneer answers back—and his response to the Gentleman's sardonic question is not to deny the connection, but rather to double down on it. "I know no Reason", he replies defensively, "why *innocent Diversion* should not be encouraged amongst me and my Correspondents, as well as amongst Children". The Gentleman, moreover, concedes the point: "I must confess", he says, "if it be agreeable, I have nothing to say against it", commenting at another point in *The*

---

[6] Shadwell, *The Virtuoso*, E1r (27). On this scene, see Tribble, "Watery Knowledge", 119–20.

[7] Shadwell, *The Virtuoso*, D3v (24). On the proverbial idea of old age as a second infancy, see Covey, "A Return to Infancy".

[8] King, *The Transactioneer*, K2v-3r (98[68]-69). The phrases in italics are quotations from Benjamin Bullivant, "Part of a Letter". On the impact of *The Transactioneer*, see Lund, "'More strange than true'".

[9] King, *The Transactioneer*, K3r (69).

Transactioneer that "we are much obliged to the Doctor, for he's a great Promoter of Philosophical and Innocent Mirth".[10] If the Society's activities are useless, as Cavendish, Shadwell, and King imply, at least they are also harmless and entertaining. As Joanna Picciotto notes, then, where "satires of the virtuoso rehearse the theme that those obsessed with viewing curiosities became curiosities themselves", they also "actually promote the notion of experimental investigation as an innocent activity, a wholesome if eccentric extension of child's play".[11]

It is a central contention of this book that the terms of Cavendish's, Shadwell's, and King's critiques are revealing not because they describe the work of the Royal Society in terms which its Fellows would have rejected outright, but rather because they capitalize on a connection between experimentalism, childishness, and play that its predecessors, members, and supporters frequently made themselves. Sometimes, certainly, the association is uneasy or pejorative: thus, Robert Boyle's repeated tendency to describe his experiments as "trifles" (i.e. trinkets or toys), is one aspect of a performance of humility which appears to admit, sincerely or otherwise, the insignificance of his work.[12] Approbatory analogies between experiment and a productive, edifying form of childish play, however, are at least as common as scornful or mocking versions of the comparison. The Transactioneer's framing of the study of nature as an "*innocent Diversion*", for instance, echoes Francis Bacon, writing in the Preface to his *Instauratio Magna* (1620). Here, Bacon cites Proverbs: "Of the sciences that contemplate nature", he announces, "the divine philosopher [Solomon] declared: *the Glory of God is to conceal a thing; but the glory of a king is to find it out*". It is, Bacon continues, "just as if the Divine Nature, delighting in an innocent friendly children's game of hide and seek chose, out of his favour and goodwill towards men, the human soul as his playmate in that game".[13] Describing natural science as a cosmic

---

[10] King, *The Transactioneer*, K3r (69), D1v (18).

[11] Picciotto, *Labours of Innocence*, 246–7.

[12] See, for example, Boyle, *Some Considerations Touching the Usefulnesse*, **4v (unpaginated), C4r (23), L2v (92), Pp3r (309); or Boyle, *Certain Physiological Essays*, F3r (37).

[13] Bacon, Preface to the *Instauratio Magna*, in *The Oxford Francis Bacon* (henceforth, *OFB*), 11.24. Bacon cites Proverbs 25:2 (KJV), as he also does in a similar passage in *The Advancement of Learning*, *OFB*, 4.36. On how these passages present "scientific investigation" as "a ludic engagement with the divine" which operates, through allegory, see Poole, "Bacon and Allegory", 119, 125. On creation as divine playfulness, see Bredekamp, "The Playfulness of Natural History", 67.

game of hide-and-seek with God, Bacon reacts against a long theological tradition which had stressed the spiritual dangers of curiosity, vindicating natural philosophy as "innocent" by figuring it as a childish frolic.[14]

Bacon's framing of childish play as innocent should, perhaps, give us pause, for it represents a conspicuous departure from broader cultural and religious attitudes, in a period when an Augustinian emphasis on original sin—and consequently, the moral depravity and mental limitations of children—tended to be prominent. As Augustine wrote in his *Confessions*, quoting Job, "*in thy sight can no man bee cleane from his sinne;* no not an Infant of a day old … it is not the minde of Infants that is harmelesse, but the weaknesse of their childish members".[15] Similarly, for John Calvin, "the uncleannesse of the parentes so passeth into the children, that all withoute anye exception at their beginninge are defiled".[16] This position was subsequently adopted by the reformed Church of England and disseminated by the *Book of Common Prayer*, with article nine of the Thirty-nine Articles of Religion (passed in 1563), affirming that "Original Sin … is the fault and corruption of the nature of every man, that naturally is ingendered of the off-spring of *Adam* … and therefore in every person born into this world, it deserveth God's wrath and damnation".[17]

The generally derogatory attitude to childhood that this theological tradition encouraged is also evident in the work of Bacon and of early Royal Society members. In his *Occasional Reflections upon Several Subjects* (1665b), for instance—a series of meditations largely written in Boyle's late teens and early twenties, though published some years later—Boyle contrasts the child's grasping, passionate pleasure in the heavenly bodies with his own more mature and measured observations:

---

[14] Notably, St Augustine calls curiosity as a "vayne and curious itch … masked under the title of *Knowledge* and *Learning*", describing it as "many wayes more dangerous" than "concupiscence"; see his *Confessions*, Gg7r-v (683–4). Numerous scholars have explored the perceived dangers of curiosity in the Christian tradition. See, *inter alia*, Daston and Park, *Wonders and the Order of Nature*, 305–16; Benedict, *Curiosity*; Harrison, "Curiosity, Forbidden Knowledge, and the Reformation of Natural Philosophy"; Harrison, *The Fall of Man*, 34–6 and *passim*; and Evans and Marr, eds., *Curiosity and Wonder*.

[15] Augustine, *Confessions*, B10r-11r. (19–21); Job 15:14–16 (KJV). On Augustine's ideas about childish (non)innocence, see Stortz, "'Where or When Was Your Child Innocent?'".

[16] Calvin, *The Institution of Christian Religion*, Book 2, A3r-v.

[17] "The Thirty-nine Articles of Religion", in Cummings, ed. *The Book of Common Prayer*, 676.

Amongst those numerous Eyes, that these fair Lights attract in so clear a night as this, there are not perhaps any that are more delighted with them, than this Child's seem to be. ... But his is a pleasure, that is not more great than unquiet, for it makes him querulous, and unruly, and because he cannot by his struggling, and reaching forth his little hands, get possession of these shining Spangles, that look so finely, their fires produce water in his eyes, and cries in his mouth ....

Whereas, though my inclinations for Astronomy make me so diligent a Gazer on the Stars, that ... I gladly spend the coldest hours of the night in contemplating them; I can yet look upon these bright Ornaments of Heaven it self, with a mind as calm and serene, as those very nights that are fittest to observe them in.[18]

Both Boyle and the child are delighted by the stars, but the boy's greedy and irrational desire to possess them, and his grief on finding he cannot, is juxtaposed with Boyle's more mature mastery of his passions in the pursuit of astronomical knowledge. As "a person that has his Affections, and Senses, at [his own] command", Boyle claims with Stoical self-satisfaction, his is not an "unquiet Pleasure" but a "rational Contentment".[19]

Elsewhere, ideas about the innate corruption and cognitive deficiencies of children inform a narrative which frames experimental philosophy not as a form of play, but rather as contiguous with the acquisition of maturity, wisdom, and authority—often in contrast to the unproductive, ignorant verbosity of the childish ancients. Complaining in the *Novum Organum* (1620) that "men have been held back from further progress in the sciences by the siren song of reverence for antiquity", Bacon effects a dizzying inversion: "The dotage and old age of the world", he asserts, "should be taken for antiquity in its right sense, and ought to denote our times, and not the springtime of the world when the ancients lived". If the distant past was the youth of the world, then in the present it has become elderly, and just "as we expect greater knowledge of human

---

[18] Boyle, *Occasional Reflections*, Ll6r-v (171–2).

[19] Boyle, *Occasional Reflections*, Ll7r-v (173–4). Boyle also contrasts the naïve, passionate apprehension of a child with the skill and self-control of the natural philosopher in *Some Considerations Touching the Usefulness*, B2r (4): "The Book of Nature is to an ordinary Gazer, and a Naturalist, like a rare Book of Hieroglyphicks to a Child, and a Philosopher: the one is sufficiently pleas'd with the Odnesse and Variety of the Curious Pictures that adorne it; whereas the other is not only delighted ... but receives a much higher satisfaction in admiring the knowledg of the Author, and in finding out and inriching himselfe with those abstruse and vailed Truths dexterously hinted in them".

affairs and maturity of judgement from an old man than from a youth, because of his experience ... so in the same way much greater things could reasonably be expected from our time ... than from the earliest ages".[20] Correspondingly, the Greek philosophers who had for so long been taken as authorities on nature are "forever children. ... And they have indeed one attribute of children, i.e. they are quite ready to prattle but cannot procreate. For their wisdom seems to be full of words but barren of works".[21] For Bacon, moreover, in their reliance on past authorities the scholastics were no better: Bacon describes the scholastic theory that "dense and solid bodies are carried towards the centre of the Earth" as "a silly and childish idea".[22]

Bacon's later admirers echoed and expanded on this kind of rhetoric. According to the pamphleteer and physician Marchamont Nedham, writing in his *Medela Medicinæ* (1665), "the reputed Oracles of our Profession, *Hippocrates* and *Galen* ... [were] children in the Art, such as lived in the nonage of true Philosophy and Physick". The Hippocratic doctrine of critical days (days on which a disease reaches crisis point), in particular, was "as childish a Conceit as ever was owned by any Long Beards", comparable to play in its reliance on random chance: "for, truly it is like the Children's game called *Ludere Par Impar, Even and odd*".[23] The physician and chymist John Webster, meanwhile, gives this way of thinking a gendered inflection in his 1654 *Academiarum Examen*, describing scholastic academic exercises as "full of childishness", and celebrating instead the manly, martial maturity of the advocates of experimental science. These "valiant champions" who "have stood up to maintain truth against the impetuous torrent of antiquity, authority and universality of opinion", he pontificates, "are ... no babes, but strong men, who fight not with the plumbeous weapons of notions, *Syllogism,* and [dis]putation, but with the steely instruments of demonstration, observation, and experimental induction".[24] Similarly, in his poem "To the Royal Society", first published in Thomas Sprat's *The History of the Royal-Society of London* (1667), the Royalist poet and essayist Abraham Cowley invents a coming-of-age story

---

[20] Bacon, *Novum Organum, OFB,* 11.133.

[21] Bacon, *Novum Organum,* 11.115; see also the Preface to the *Instauratio Magna,* 11.12. Rees and Wakely note that in his assertion that the Greeks are "forever children", Bacon quotes the Egyptian priest in Plato's *Timaeus,* 22b (*OFB,* 11.606).

[22] Bacon, *Novum Organum,* 11.318.

[23] Nedham, *Medela Medicinæ,* Z1r (337), X1v-2r (306–7).

[24] Webster, *Academiarum Examen,* O2v (92), B2v (unpaginated).

for a personified and masculinized "Philosophy" which has until recently "bin kept in Nonage" by the scholastic "Guardians" who, fearing that Philosophy's "Natural Powers" would "put an end to their Autoritie", spoon-fed him with the "Desserts of Poetry ... Instead of solid meats t'encreas his force". From this state of artificial, infantilizing servitude, Philosophy is liberated by "a mighty Man": Bacon himself, who, Cowley declares, "boldly undertook the injur'd Pupils caus", restoring to him to the privileges of his majority.[25]

In these examples, Bacon, Nedham, Webster, and Boyle differ from Cavendish, Shadwell, and King in their insistence that it is the classical authorities and medieval Schoolmen who should be understood as childish and unproductive, not the champions of the new experimental philosophy (who are, rather, men in their prime). At the same time, they share with experimentalism's critics an assumption that childhood is a passionate but unproductive state characterized by error, indignity, and imbecilic belligerence, which should ideally be speedily discarded.

In spite of their enormous influence, however, such ideas were not undisputed: whether or not we date the emergence of the idea of childhood innocence to the seventeenth century, as some historians have suggested we should, more positive evaluations of this life stage were not uncommon in this period.[26] The Catholic church had maintained that a child was incapable of mortal sin until the age of seven, and a long theological tradition, running from Clement of Alexandria to Julian of Norwich to Dante, took a more favourable view of children as innocent and/or possessed of preternatural knowledge.[27] As Dante wrote in *The Divine Comedy*, "in little children only mayst thou seek/True innocence and

[25] Cowley, "To the Royal Society", in Sprat, *The History of the Royal Society*, B1r-v. Elsewhere Cowley celebrates "the Industry of Men" who "work amongst Gods creatures, instead of Playing among their Own". Cowley, *A Proposition*, A6r (preface; unpaginated).

[26] On the idea that the modern idea of childish innocence arose in the seventeenth century, see Higonnet, *Pictures of Innocence*, 8.

[27] See Wall, "Childhood in Western Philosophy"; and Davis, "Brilliance of a Fire", 383. See also Shulamith Shahar, who explores how in the Middle Ages "due to ... his innocence, the child is sometimes privileged to grasp a significant truth which is still hidden from adults". Shahar, *Childhood in the Middle Ages*, 17–18. In early modern England, Shakespeare's preternaturally precocious youths—from the young Henry VI in *1 Henry VI* (c.1591; 1623) to Mamillus in *The Winter's Tale* (c.1611; 1623)—seem to embody this tradition in their sharp insights and sententious wisdom. See, for instance, Henry's political moralizing: "Believe me, lords, my tender years can tell/Civil dissension is a viperous worm /That gnaws the bowels of the commonwealth". Shakespeare, *King Henry VI, Part 1*, 3.1.71–3.

faith".[28] This tradition was not easily dismissed by reformers: scriptural legitimation for the more approbatory attitude to childhood could be found in the Gospel of Matthew, where Christ calls a "little child" before his disciples, urging them to "be converted, and become as little children", in order to "enter into the kingdom of heaven".[29] And indeed, as Robert Davis has argued, in the early modern period the concept of childhood innocence is pervasive, finding "its most defiant assertion" in (but certainly not limited to) the works of seventeenth-century Puritans and dissenters, the literature of religious mysticism, and the esoteric tradition.[30] In the words of the cleric Jeremy Taylor in his 1655 *Unum Necessarium, or the Doctrine and Practice of Repentance*:

> It is hard ... to reckon all children to be born enemies of God ... full of sin and vile corruption when the Holy Scriptures propound children as imitable for their pretty innocence and sweetness, and declare them rather heirs of heaven than hell ... [hell] was 'prepared for the devil and his angels' not for innocent babes. This does not call them naturally wicked, but rather naturally Innocent.[31]

Here, childish innocence is held up as exemplary and childhood is celebrated as a state of grace and of superior insight.

Despite challenges from alternative theories, moreover, when it came to the important question of the origins of the human soul the reformed Church held steadfastly to the doctrine known as creationism: that is, the idea that each human soul is formed individually and *ex nihilo* by God and infused into the foetus' or infant's body during conception, gestation, or birth.[32] Confronted with the knotty problem of reconciling the soul's

---

[28] Dante, *The Divine Comedy*, 229; quoted in Davis, "Brilliance of a Fire", 383.

[29] Matthew 18:2–4 (KJV).

[30] Davis, "Brilliance of a Fire", 383.

[31] Taylor, *Unum Necessarium*, Ee8r-v (399–400); also cited in Davis, "Brilliance of Fire", 383.

[32] The most notable alternative theories were traducianism (the idea, originating with Tertullian's *Treatise on the Soul* [c.197–220 A.D] and revivified by Luther in the sixteenth century, that souls are generated with the body and are infused seminally with original sin) and pre-existence (the idea, which has its roots in Platonic philosophy but which was most fully expounded by Origen of Alexandria, that all souls were created by God at the beginning of time and only subsequently fell into embodiment). On the three main theories regarding the origins of the soul, see Allen, *Doubt's Boundless Sea*, 159–62; Lewis, "Of 'Origenian Platonisme'", 271; Almond, "The Journey of the Soul", 778.

divine origin with its inheritance of original sin—for if human souls are made by God, surely they should be purely good?—theologians came up with a range of solutions.[33] Calvin, for instance, both argued that "oure destruction commeth of the faulte of oure own fleshe & not of God" and asserted that although "infantes ... have not as yet brought forth the fruytes of theyr owne infirmitie, yet they have the seede thereof enclosed within them".[34] If, then, original sin is still a "seede" in the soul of infants, children may have a distinct moral and epistemological advantage over adults, retaining vestiges, at least, of a prelapsarian purity.[35] This tradition also left its mark on early modern culture—including the work of Bacon and his self-proclaimed followers in the Royal Society, where creationism also remained prevalent and where, as this book will show, childhood is also often depicted in highly positive terms.[36]

Over the past few decades, researchers have done much to move the historiography of sixteenth- and seventeenth-century "science" away from an older model which celebrated the revolutionary contributions of a few great men, such as Bacon and Isaac Newton, towards a more nuanced understanding of the multiple contributions made by (or coerced from) women, the emergent middle and working classes, and people from a range of racial and ethnic groups (including enslaved people). In recent years, for instance, scholars including Sarah Hutton, Elaine Leong, and

[33] Notably, Augustine agonized about how to reconcile creationism with original sin in his letters to Jerome; see *A Treatise on the Origin of the Human Soul*. For a discussion, see Kelly, *Early Christian Doctrines*, 344–56.

[34] Calvin, *Institution of Christian Religion*, Book 2, A4v-5r (4–5).

[35] Richard Baxter expands Calvin's horticultural image, comparing "original sin" to "the arched Indian Fig-tree, whose branches turning downwards and taking root, do all become as trees themselves". In children, however, this tree it is yet "but *a* twig". Baxter, *A Christian Directory*, Uuu2r (513).

[36] Boyle, for one, affirmed creationism, writing that "the Origine of [the] Immortal Soul ... was Gods own immediate Workmanship, and was united to the Body already form'd". Boyle, *The Excellency of Theology*, C8r (25). On the Royal Society's antipathy to pre-existence, as advocated by one of its members, Joseph Glanville, see Lewis, "Of 'Origenian Platonisme'". Bacon himself took a nuanced position: in his *De Augmentis Scientiarum* (1623)—a Latinized and expanded version of *The Advancement*, re-translated into English in its expanded form in 1640—he offered a third way between creationism and traducianism. Asserting that the human soul "*hath two Parts*", namely the reasonable soul, which is "*a thing Divine*", and the "*unreasonable*" or "*sensitive soul*", which "*is common to us with Beasts*", Bacon contends that the first "hath it's originall from the *Breath of God*" whilst the latter comes from "the *Matrices of the Elements*". The rational soul is divinely created and infused; the sensitive soul is "traduced". Bacon, *Of the Advancement*, Cc2v (206).

Wendy Wall have drawn attention to the central importance of women's domestic expertise to early modern experimental cultures; Pamela H. Smith and Deborah E. Harkness have enriched our knowledge of the contributions made by artisans, craftspeople, other manual workers, and merchants; and Londa Schiebinger, Susan Scott Parrish, and Pablo F. Gómez, amongst others, have highlighted the necessity of understanding how early modern science developed in tandem with colonial expansion, often drawing on indigenous knowledges or exploiting slave labour.[37] By contrast, children have received scant attention.

Why? For a start, the narratives of modernity as a process of disenchantment, which—following Max Horkheimer and Theodor W. Adorno's *Dialectic of Enlightenment* (1947)—dominated much twentieth-century historiography, made Francis Bacon into a figurehead for Enlightenment efforts to demystify and dominate the natural world. As what Joe Moshenska resonantly calls the "doyen of disenchantment", Bacon is a grimly unappealing, "dourly joyless" character whose instrumentalization of reason is profoundly inimical to the apparent frivolity and evident pleasurableness of child's play.[38] Similarly—and despite offering a more generally optimistic account of Enlightenment rationality than Horkheimer and Adorno—Hans Blumenberg's *The Legitimacy of the Modern Age* (originally published in German in 1966) frames Baconian science as dependent on the expulsion of "nature's unseriousness … its supposed playfulness" in favour of a new emphasis on "nature's serious usefulness".[39] Perhaps, too, the narrative influentially propounded by Steven Shapin, which ties the emergence of English experimental philosophy to the legitimating force of gentlemanly civility and self-discipline, has contributed to this disregard of children.[40] If, as Shapin suggests, early scientific culture secured credibility

---

[37] For the contributions of women, see, Hunter and Hutton, eds., *Women, Science and Medicine*; Hutton, "Science and Natural Philosophy"; Wall, *Recipes for Thought*; and Leong, *Recipes and Everyday Knowledge*. On artisans and merchants, see Smith, *The Body of the Artisan*, and Harkness, *The Jewel House*. On indigenous knowledges, see Schiebinger, *Plants and Empire*; Parrish, *American Curiosity*; and Gómez, *The Experiential Caribbean*.

[38] Moshenska, *Iconoclasm as Child's Play*, 180.

[39] Blumenberg, *The Legitimacy of the Modern Age*, 475. Moshenska discusses Blumenberg's suggestion that a perception of seriousness or playfulness is "one of the important ways in which an epochal threshold is defined" in *Iconoclasm as Child's Play*, 182–3.

[40] On the importation of "the conventions, codes, and values of gentlemanly conversation into the domain of natural philosophy", see especially Shapin, *A Social History of Truth*, xvii, and *passim*.

by an appeal to its practitioners' possession of a complex blend of gentle-manly qualities including economic independence, the possession of moral virtues such as prudence (a form of wisdom deriving from accumulated experience), and genteel social conduct, and it is no wonder children—economically dependent, often rash and unreasonable, socially unre-fined—have been overlooked.[41]

That is not to say that children, or play, have been ignored entirely. In his *Homo Ludens*, first published in 1938, the Dutch historian and great theorist of play Johan Huizinga resists the "conclusion … that all science is merely a game", but comments that, nonetheless, "in earlier times and right up to the Renaissance … scientific thought and method showed unmistakable play-characteristics".[42] More recently, Catherine Wilson has remarked on "the positive valuation, in the educational literature from John Amos Comenius to John Locke, of children's naïve curiosity and delight in novel experience. … The charm of what Hooke calls the 'real, the mechanical, the experimental philosophy' lies in its similarity to child's play"—a claim subsequently taken up by Joe Moshenska, in his book *Iconoclasm as Child's Play*.[43] Urging us to question those "large-scale nar-ratives that see play as depleted or disappearing under modernity", Moshenska comments that in the seventeenth century, "play might be a way of … knowing and engaging" with "the divinely formed world", and that as such "in natural philosophical as in pious forms of activity, play keeps recurring, not as the opposite of valuable and worthwhile pursuits but as a version of them".[44] Focusing on eighteenth-century Prussia, meanwhile, Kelly Joan Whitmer has illuminated the practical, concrete contributions made to the understanding of "natural processes" and the refinement of "observational procedures" by the "young people" of the Halle Orphanage: a Pietist institution which, Whitmer shows, also func-tioned as a "scientific community".[45]

---

[41] On the importance of the mature virtue of "prudence", defined in the early modern period as the wisdom to act "on the basis of accumulated past experience", see Shapin, *A Social History of Truth*, 238–9.

[42] Huizinga, *Homo Ludens*, 203–4; see also Raymo, "Science as Play".

[43] Wilson, *The Invisible World*, 23–4. See also Keith Thomas, who comments in passing that "the new science, with its hostility to the ancients, was to produce the dogma than only the young could make intellectual discoveries" in "Age and Authority", 246–7.

[44] Moshenska, *Iconoclasm as Child's Play*, 181–2.

[45] Whitmer, *The Halle Orphanage*, 2–3.

Perhaps most consequentially, Paula Findlen—whose work is also discussed in Chap. 5 —has done much to illuminate how the humanist tradition of *serio ludere* (playing seriously) informed the work of some
Renaissance natural philosophers, from the myriad natural historians for
whom all kinds of inexplicable or unusual phenomena were designated
lusus naturae, nature's jokes, to mathematicians and astronomers including Johannes Kepler, for whom "play ... revealed the pattern of the universe, conveyed by God through the artistry of nature".[46] It is significant,
however, that in Findlen's account play only becomes useful as a means of
pursuing knowledge once its associations with childhood are elided or
suppressed: in order to be viewed as "a divine activity" with the potential
to uncover the secrets of nature, "games" must be elevated above "merely
child's play".[47] According to Findlen's earlier work, moreover, the ascen-
dancy in the seventeenth century of the new experimental and mathematical philosophies led to an expulsion of play, as first Bacon and then "the
early Royal Society members attempted to purge their interpretations of
nature of any traces of the ludic, which they associated with paganism,
atheism, and an excessive love of rhetoric".[48]

In contrast, in this short book I argue that children played a crucial role
in the development of early experimental science in seventeenth-century
England, both as rhetorical exemplars of the kinds of attributes to which
adult experimentalists should aspire and as active (and playful) participants
in their activities. Bacon and his devotees in the Royal Society made much
of their aristocratic status, to be sure, but the form of empiricism pro-
moted by them grounds knowledge not in elite learning, but rather in a
form of sensory attentiveness potentially accessible not only to people of
all classes and sexes but also to people of all ages. As Robert Hooke writes
in the preface to his *Micrographia* (1665), the "*reformation* in Philosophy"
desired by the Royal Society does not require any "exactness of *Method*, or
depth of *Contemplation*", but only "a *sincere Hand*, and a *faithful Eye*, to

---

[46] Findlen, "Between Carnival and Lent", 257; see also Findlen, "Jokes of Nature", and
Findlen, "Ludic Postscript".

[47] Findlen, "Between Carnival and Lent", 255. Moshenska similarly notes that where "play
is salvaged and described as far from a trifling matter" by Kepler and others, its valorization
depends on its distance "from the material reality of children and their playthings". *Iconoclasm
as Child's Play*, 31.

[48] Findlen, "Between Carnival and Lent", 262–3. Findlen nuances this chronological narrative in a later piece, "Ludic Postscript", which explores the endurance of *lusus naturae* in
various forms into the twenty-first century.

examine ... the things themselves as they appear".[49] Similarly, Robert Boyle declares in the Preface to *Some Considerations About the Reconcileableness of Reason and Religion* (1675) that although "ratiocinations ... depend, in some measure, upon the judgment and skill of those that make the Observations whereon they are grounded":

> There are some Arguments, which being clearly built upon Sense or evident Experiments ... may, I think, be fitly enough compar'd to Arrows shot out of a Cross-Bow, or Bullets shot out of a Gun, which have the same strength, and pierce equally, whether they be discharg'd by a Child, or a strong Man.[50]

Forms of knowledge built upon sensory experience or "Experiments", Boyle suggests here, may be produced with equal effectiveness by "a Child, or a ... Man", regardless of age.

In fact, Chap. 2 proposes, in the works of some figures associated with the Royal Society, children are thought to have specific sensory and mental characteristics which make them consummate empiricists. In particular, I suggest, aspects of children's perceptual and cognitive experience which were, in the rationalist philosophy associated with René Descartes, considered as barriers to the acquisition of truth—namely their intense sensory impressionability, their receptive memories, and their inability to engage in high-level abstract reasoning—could be advantages in the context of a methodological programme which professed to value first-hand experience above all else, and which shrunk from theoretical conjecture, premature systematization, and dogma. For all that, the distinction is not as clear-cut as we might be inclined to think: aspects of Descartes' thought also imply that, in some regards, infancy is an epistemologically privileged state. The final part of the chapter considers how ideas about the benefits of a childlike disposition endured into the eighteenth century in the work of the "common sense philosopher" Thomas Reid.

Chapter 3 explores connections between ideas about childish innocence and the emergence in this period of something like the modern scientific virtue of objectivity, as well as the implications of this link for our understanding of the historical development of the latter concept. Defined partly as freedom from worldly prejudice and preconception, as well as a

---

[49] Hooke, *Micrographia*, a2v (unpaginated). See also Bacon, *Novum Organum*, 11.97: "My view of the process of discovering the sciences is this: that little be left to sharpness and force of wits, but that wits and intellects be put on much the same footing".

[50] Boyle, *Some Considerations About the Reconcileableness of Reason and Religion*, a3r (xvii).

state of prelapsarian moral perfection, childlike innocence was increasingly held up as an epistemological as well as an ethical ideal, facilitating perceptual lucidity and accuracy. In particular, childish ignorance of worldly standards of value was both used to satirize the indiscriminate collections of seventeenth-century philosophers, and as evidence for their innate insight and piety in preferring divinely created things over the bogus glitter of earthly treasures. Ancient ideas about the autotelic character of play were also important in this regard, allowing thinkers to postulate a form of experiment which was productive precisely insofar as it was purposeless.

Chapter 4 investigates the ways in which natural philosophers including Robert Hooke attempted to emulate or recreate a childlike perspective in their own experimental activities. For some members of the early Royal Society, both the playfulness and the sensory lucidity of childhood could be partially recaptured in adulthood using prosthetic aids. In particular, the telescope and microscope were—for all their technological complexity— sometimes understood in a very literal way as toys which nonetheless served a vital function, promising to restore something of this lost childlike sensibility. This effort, I suggest, was underpinned by a transformation in ideas about experience as a source of knowledge, as an Aristotelian model of "experience" as conferring understanding only gradually, over an extended period of time (and hence, as the exclusive preserve of the old), was replaced by a new, Baconian emphasis on the heuristic and probative value of experimental intervention in the present moment (a form of experience at least equally available to children).

In some contexts, then, experiment was considered a form of childish play in the seventeenth century. The relation, however, was isomorphic: play, as Chap. 5 argues, was also considered a form of experiment. The pervasive framing of the natural sciences as play in the work of seventeenth-century experimental philosophers, moreover, is not merely rhetorical hot air: actual, real-life children were active participants in the natural philosophical world of seventeenth-century England, as children's games as treated as a kind of spontaneous, *ad hoc* experimentation from which quite sophisticated theories about the natural world might eventually develop. Notably, bubble-blowing played a critical role in the development both of Francis Bacon's ideas about the nature of fluids and of Isaac Newton's ground-breaking theories of light and colour.

Finally, a short Conclusion to the book explores the endurance of seventeenth-century ideas about the childishness and science into the eighteenth century and beyond, speculating that in insisting on the isomorphism of experiment and play, members of the early Royal Society contributed to a reconfiguration of ideas about childhood which still informs how we think about youth and education today.

At this point, it is worth acknowledging that the question of how "childhood" was defined and understood in seventeenth-century England is a vexed one. Philippe Ariès' claim, in his *Centuries of Childhood*, that "in medieval society the idea of childhood did not exist"—subsequently taken up in relation to sixteenth- and seventeenth-century England by the historian Lawrence Stone, amongst others—has been roundly refuted by scholars including Linda Pollock and Nicholas Orme, and there now is a general consensus that in the early modern period childhood was recognized a distinct life stage, conceptualized in ways which were in some regards (though certainly not all) continuous with how we think of it today.[51] In the present context, the presence of a concept of childish innocence in the early modern period, discussed above, is one important example of this continuity. So, too, is a widespread emphasis on children's preference for play over laborious book-learning: witness the Puritan theologian and moralist Thomas Goodwin, who, in *The Vanity of Thoughts Discovered* (1638), disapprovingly describes the reluctance of "Schoole-boyes ... to goe to their Books ... their heads being full of play".[52]

At the same time, there were significant differences to ideas about and in the experience of childhood in this period: many children, for example, took on more adult responsibilities in the household than they usually do today.[53] The boundaries of childhood were also fluid and contested: whilst influential, the ancient model of the *aetates hominum* or ages of man, formalized in the Middle Ages, left considerable scope for interpretation.[54] According to the influential twelfth-century encyclopaedist Angelicus

[51] Ariès, *Centuries of Childhood*, 125; according to Ariès, children were not considered extremely vulnerable or sexually innocent, as they are today; rather, they tended to be considered as adults "reduced to a smaller scale" (33). See also Stone, *The Family, Sex and Marriage*. For refutations, see Pollock, *Forgotten Children*, and Orme, *Medieval Children*.

[52] Goodwin, *The Vanity of Thoughts*, B11r (37). See also Roger L'Estrange: "he that will be Angry with *Any Man*, must be displeas'd with *All*; which were as ridiculous, as to quarrel with ... a Schoolboy for loving his *Play* better than his Book". L'Estrange, *Senecas Morals*, E4v (56). In Greek, the verb "to play" (*paizō*) is rooted in the noun "child" (*pais*), so to play is literally something like "to child-ize". See Kidd, *Play and Aesthetics*, 20.

[53] Ferraro, "Childhood", 64–5.

[54] On the fluidity of "childhood" as a category, see, for instance, Lamb, *Reading Children*, 12–13. On the formalization of the *aetates hominum*/ages of man scheme in the Middle Ages, see Witmore, *Pretty Creatures*, 27–8. See also Ariès, *Centuries of Childhood*, Chapter 1.

Batholomaeus, whose work was translated into English in the fourteenth century and reproduced a number of times thereafter, the first stage of childhood—defined by the child's inability to "speake nor sound his words perfectly"—ended at seven, with the "second age" of childhood, or "*Puericia*", occurring between seven and fourteen years, and "*Adolescentia*" occupying an indefinite duration between fourteen years, and either twenty-one, twenty-eight, thirty, or thirty-five years.[55] By contrast, for Henry Cuff, writing in *The Differences of the Ages of Man's Life* (1607), childhood lasted until the age of twenty-five—around the period when an apprentice would complete his training—and was more precisely divided into infancy (from birth to around three or four years) and boyhood (ending around nine or ten), followed by "our budding and blossoming age" and "youth".[56] Later in the seventeenth century, the educational theorist John Dury, a close associate of Samuel Hartlib and (with Robert Boyle) a member of the so-called Invisible College of men who gathered to discuss experimental ideas in London in 1640s, reckoned that childhood proper lasted until around thirteen or fourteen, but was followed by a period of "youth" lasting till "nineteen or twenty".[57]

Given this diversity, determining the exact numerical boundaries of childhood in this period is neither feasible nor—for my purposes here—necessary, for this is not a book about children themselves. I am not, that is, concerned either with determining how childhood was experienced in this period or with identifying a dominant cohesive "view" of childhood. Rather, this book explores the ways in which multiple and shifting notions of and perceptions of childhood—including ideas about their innocence, sensory receptivity, curiosity, credulity, indiscrimination, unpredictability, and playfulness—were mobilized to a range of rhetorical and practical ends by seventeenth-century natural philosophers as they strove to conceptualize and valorize their own methods. As such, I take a flexible and responsive approach, allowing the authors discussed to themselves arbitrate the limits and outline the contours of this most opaque and labile stage of life.

*   *   *

[55] Bartholomaeus, *Batman uppon Bartholome*, N4v (70).
[56] Cuff, *The Differences*, I2r-4r (115–19).
[57] Dury, *The Reformed School*, C2r-v (51–2).

On 30 May 1667, Margaret Cavendish attended a meeting of the men she had mocked a year earlier in the *Observations*, an event at which Samuel Pepys, as a fellow of the Royal Society, was also present, and which he subsequently described in his diary on 30 May 1667. "The Duchesse", Pepys noted,

> hath been a good comely woman; but her dress so antic and her deportment so unordinary, that I do not like her at all, nor did I hear her say anything that was worth hearing, but that she was full of admiration, all admiration. Several fine experiments were shown her of Colours, Loadstones, Microscope, and of liquors. ... Here was Mr Moore of Cambrige, whom I had not seen before, and I was glad to see him—as also a very pretty black boy that run up and down the room, somebody's child in Arundell-house. After they had shown her many experiments ... she departed, being led out and in by several Lords that were there; among others Lord George Barkely and Earl of Carlisle and a very pretty young man, the Duke of Somersett.[58]

Several scholars have explored this account from a feminist perspective.[59] Most pertinently here, Lisa T. Sarasohn argues that Cavendish orchestrated the visit as assertion of female independence. Cavendish, Sarasohn proposes, deliberately (and playfully) made a spectacle of herself through her eccentric clothing and demeanour in order to undermine "the dogged seriousness of the Royal Society", thereby highlighting "the performative nature of experimental science" and making her distinguished hosts seem like "sideshow entertainers". For all her outspoken "admiration" (a term which in this period could in any case indicate mere astonishment, rather than appreciation), then, her visit was an extension of, rather than a departure from, the critique of the *Observations*, restating in another form her conviction that the Royal Society's activities were no more than "play and display". As such, Cavendish—despite her sex—"could reinforce her claim to be the equal or even superior of her fellow investigators of nature".[60]

---

[58] Pepys, *Diary*, 8.243.

[59] Ryan J. Stark, for instance, comments that "Pepys' complaints about the 'pomp' surrounding Cavendish, including her clothing, are a clear effort to invoke the common trope characterizing the feminine mind as preoccupied with flattery and ornament, not reason and serious science". Stark, "Margaret Cavendish and Composition Style", 276.

[60] Sarasohn, *The Natural Philosophy of Margaret Cavendish*, 28.

Less frequently noticed than the proto-feminist undertones of Cavendish's visit is the presence of another marginalized group: children. For a start, the "very pretty young man, the Duke of Somersett", who helps escort Cavendish out, William Seymour, was born in April 1652, making him just fifteen on the occasion described—and, therefore, if no longer quite a child, certainly not a fully grown man. In addition to the fresh-faced Duke, moreover, Pepys comments approvingly on the presence of "a very pretty black boy" who "run[s] up and down the room".[61] Offering a more impulsive, unselfconscious version of Cavendish's strategic, theatrical form of playing, the presence of the boy, running aimlessly and exuberantly to and fro, introduces into the scene precisely the note of unfocused, undisciplined, "unprofitable" energy she objected to in the *Observations*. What happens, this book asks, if we track his footsteps and those of his peers through both the rooms of early modern experiment, and through the minds of the men who inhabited those places?

---

[61] Here, "black" probably refers to the boy's hair and eye colour; "Black (*adj.*), sense I.2.a", *Oxford English Dictionary*, s.v. June 2024, https://doi.org/10.1093/OED/1350335278. In the same entry Pepys describes Elizabeth Ferrabosco, a woman of Italian origin, as "black", commenting on her "good black little eyes".

CHAPTER 2

"The tenderest touch": Children's Senses

Abstract This chapter proposes that in the works of some figures associated with the Royal Society, children are thought to have specific sensory and mental characteristics which make them consummate empiricists. In particular, I suggest, aspects of children's perceptual and cognitive experience which were, in the rationalist philosophy associated with René Descartes, considered as barriers to the acquisition of truth—namely their intense sensory impressionability, their receptive memories, and their inability to engage in high-level abstract reasoning—could be advantages in the context of a methodological programme which professed to value first-hand experience above all else, and which shrunk from theoretical conjecture, premature systematization, and dogma. For all that, the distinction is not as clear-cut as we might be inclined to think: aspects of Descartes' thought also imply that, in some regards, infancy is an epistemologically privileged state. The final part of the chapter considers how ideas about the benefits of a childlike disposition endured into the eighteenth century in the work of the "common sense philosopher" Thomas Reid.

Keywords Children • Science • Natural philosophy • Experiment • History of the senses • Empiricism • botany • Royal Society • Thomas Sprat • Francis Bacon • John Ray • Carl Linnaeus • René Descartes • Common Sense Philosophy • Thomas Reid

E. L. Swann, *Science as Child's Play in Seventeenth-Century England*, https://doi.org/10.1007/978-3-031-75849-2_2

21

"Children", declares the Catholic priest and writer Thomas Wright in *The Passions of the Mind* (1601), "lacke the use of reason" and "are guided by ... nothing else but that pleaseth their sences, even after the same maner as bruite beastes doe".[1] Thirty years later, the poet and moralist Richard Brathwaite strikes a similar note in his conduct book *The English Gentleman* (1631): "So exposed is *Youth* to *sense*", he sniffs, "and so much estranged from *the* government of *reason*; as it prosecutes with eagernesse whatsoever is once entertained with affection".[2] Wright's and Brathwaite's words can stand as representative: reflecting the influence of ancient authorities including Plato, Aristotle, and Galen, as well as the Augustinian emphasis on original sin, children in early modern England were often understood as fundamentally irrational, unable to engage in high-level abstract thinking, and immersed instead in the pursuit of physical pleasures and passions.[3] Creatures of body rather than mind, sensation rather than ratiocination, children for Wright and many of his peers were both spiritually corrupt and cognitively deficient: little more than animals.

Later in the seventeenth century, however, we find childish sensuousness appraised very differently, as containing pedagogical and even heuristic value as a means of knowing the world—a shift that chimes with a more wholesale revaluation of the epistemic potential of the senses, evident in the Royal Society's ambivalent attitude to abstract speculation and authority, and their methodological commitment (volubly proclaimed, if not always adhered to in practice) to first-hand experience.[4] In his textbook for children, *Orbis Sensualium Pictus* (1659), the Moravian educational reformer Jan Amos Comenius—whose adaptation of Baconian ideas subsequently exerted an enormous influence on pedagogical philosophies and practices—explains his inclusion of illustrations, commenting that "the senses [are] the main guides of Child-hood, because therein the Mind

[1] *Wright, The Passions of the Mind*, B4r (7). On the extent to which Wright's work (particularly its treatment of the embodied passions) should be understood as representative of the early modern period, see Sullivan, "The Passions".

[2] Brathwaite, *The English Gentleman*, E2r (27).

[3] On the ancient Greek origins of the idea that children lack reason—and the physiological underpinnings of this belief—see Kidd, *Play and Aesthetics*, 21–29.

[4] The extent to which reality reflected rhetoric in this regard is, of course, up for debate: as scholars have shown, speculation and learned authority continued to play an important role in the work of the early Royal Society, and experimental experience was highly mediated in a number of ways. Steven Shapin, for instance, emphasizes the central importance of testimony and trust, as well as "the individual's sensory confrontation with the world", to the production of scientific "truth"; see *A Social History of Truth*, 202 and *passim*.

doth not yet raise up it self to an abstracted contemplation of things".[5] For Comenius as for Wright, children are incapable of "abstracted contemplation" and instead are "guided" by their senses. The guidance in question, however, is benevolent supervision rather than brutish enthrallment, and it may be exploited for educational ends.

If, moreover, natural knowledge should "not be grounded on matters of speculation, or opinion, but onely of sence", as Thomas Sprat wrote in his *History of the Royal Society* (1667), then the child's immersion in the world of the senses is not only an opportunity for the teachers who wish them to *assimilate* natural knowledge; it also makes them, potentially, effective *producers* of such knowledge.[6] In this context, Sprat "venture[s] to propose ... whether it were not as profitable to apply the eyes, and the hands of Children, to see, and to touch all the several kinds of *sensible things*, as to oblige them to learn, and remember the difficult *Doctrines* of general *Arts?*"[7] Similarly, he grumbles, "we load the minds of Children with Doctrines, and Praecepts, to apprehend which they are most unfit, by reason of the weakness of their understandings", reiterating that "they might with more profit be exercis'd in the consideration of *visible* and *sensible things*; of whose impressions they are most capable, because of the strength of their *Memories*, and the perfection of their *Senses*".[8] Here, children's inability to engage in high-level reasoning and their reliance on their senses is no longer an impediment, but a benefit which makes them archetypal observers and experimenters: quintessential natural historians, if not philosophers capable of reasoning about causes.

Importantly, as Sprat hints in his description of the "strength" of youthful memory and the "perfection" of youthful senses, children's sensory experiences and cognitive capacities were also considered distinct from those of adults, with specific characteristics which made them especially valuable as a source of knowledge. In recent decades, sensory historians and anthropologists have become increasingly attentive to the ways that vectors such as race, gender, and class infuse and shape sensation. Assuming the cultural and historical specificity of sensory perception, such scholars also explore what Constance Classen and David Howes call "intracultural variation", acknowledging that "there are typically persons or groups who

---

[5] Comenius, *Orbis Sensualium Pictus*, A4r (unpaginated).
[6] Sprat, *History of the Royal Society*, M2v (92).
[7] Sprat, *History of the Royal Society*, Tt1r (329).
[8] Sprat, *History of the Royal Society*, Tt1v-2r (330–31).

differ on the sensory values [and practices] embraced by the society at large".[9] To take two fine examples almost at random from a wealth of possibilities, Laura Gowing's *Common Bodies: Women, Touch, and Power in Seventeenth-Century England* explores how women's experiences of (and ability to exercise) touch have been shaped by factors such as social status and reputation, as well as gender, whilst Sachi Sekimoto's and Christopher Brown's edited collection *Race and the Senses: The Felt Politics of Racial Embodiment* explores both how "multiple senses are engaged to feel race and racial differences", and "how such embodied multisensory feelings are integral to the social, political, and ideological construction of race".[10] The specificity of children's sensory experiences at particular times and in particular places, however, is relatively understudied. Building on the insight that the senses are "made, not given" (in Howes' words), historical and anthropological studies of the senses have tended to treat childhood not as a distinct life stage in itself, with its own sensory environment and capabilities, but rather as a preliminary stage in the shaping, education, or enculturation of adult perceptual habits.[11]

Early modern thinkers, on the other hand, were deeply interested in what they believed to be the specificity of the childish sensorium. As Michel de Montaigne writes in *An Apology of Raymond Sebond*, first published in 1580, "A yong childe heareth, seeth, and tasteth otherwise by natures ordinary rule, then a man of thirtie yeares; and he otherwise then another of threescore. The senses are to some more obscure and dimme, and to some more open and quicke".[12] This statement is not just an expression of Montaigne's own sceptical relativism, but of widespread

[9] Such individuals may "resist, instead of conform to, the prevailing sensory regime". Howes and Classen, *Ways of Sensing*, 12.

[10] Gowing, *Common Bodies*; Sekimoto and Brown, "Introduction", 1.

[11] Howes, "The Misperception of the Environment", 447. For an example of work focusing on the education and enculturation of the adult senses, see Harris, *A Sensory Education*. The methodological difficulties of accessing children's senses is presumably one factor in their scholarly neglect: if sensory history as a whole is heavily reliant on textual and visual sources for its recreation of the multi-sensory and synaesthetic worlds of the past, the problem is exacerbated when it comes to studying people who left little in the way of such records, leaving scholars dependent on adult recollections of and reflections on sensing in childhood.

[12] Montaigne, "An Apologie", in *Essays*, Gg1v (338). For a discussion of the specificity of youthful vision in *A Midsummer Night's Dream* and *Titus Andronicus*—and of how some Shakespeare film directors "invite the spectator to look like a child" (72)—see Carol Chillington Rutter, *Shakespeare and Child's Play*. Rutter sees *Midsummer* in particular as "a contest of looking strategies" whereby "child-sight" is characterised as "giddy, as changeable

medical orthodoxy: in this period children were thought to have a particular humoural composition which affected their perceptual experience in ways which could be epistemologically advantageous or disadvantageous. Specifically, because children were considered sanguine—hence humourally warm and moist—they were also thought to be more than usually impressionable.[13] As Henry Cuff writes in *The Differences of the Ages of Mans Life* (1607), "in our infancie wee are fullest of moisture, our experience and sense teacheth us, for so we see infants flesh most fluid and almost off a waxen disposition, ready to receive impression of any light touch".[14]

Similar statements also abound in pedagogical literature. In *The Education of Children*, translated by Richard Sherry in 1550, for instance, Erasmus advises parents to "provide that thyne infante and yonge babe be forthewyth instructed in good learnyng, whylest hys wyt is yet voyde from tares and vices, whilest his age is tender and tractable, and his mind flexible". The reason, he continues, is that "we remember nothynge so well when we be olde, as those thynges that we learne in yonge yeres … some things … the tender age perceiveth both much more quickly, & also more esily then doth the elder".[15] In *The Scholemaster* (1570), Roger Ascham, too, implicitly links children's aptness for learning both to their freedom from "vices" and to their physical tenderness and sensory quickness. "If ever the nature of man be given at any tyme, more than other, to receive goodnes", he asserts, "it is, in innocencie of yong yeares, before, that experience of evill, have taken roote in hym. For, the pure cleane witte of a sweete yong babe, is like the newest wax, most hable to receive the best and fayrest printing".[16] Here, the comparison of an infant's "witte" to unblemished wax is drawn from Aristotle's *De anima*, where "perception" is defined as "what is capable of receiving perceptible forms without the matter, as wax receives the seal of a signet ring": an image which Ascham

as taffeta, as unsettled as a gadfly, antiauthoritarian, anarchic", but also "forgiving [and] restorative", as well as "liberating". *Shakespeare and Child's Play*, 70–71.

[13] Caroline Bicks explores the gendered implications of the fact that, according to some writers, this heat reached its zenith at what we now call puberty, counteracting the supposed female tendency to humoral coolness and endowing young girls with a unique mental agility and dynamism. See Bicks, *Cognition and Girlhood*, esp. Chapter 3.

[14] Cuff, *The Differences*, I3r (117).

[15] Erasmus, *That Chyldren Oughte to be Taught* [*The Education of Children*], G1r (unpaginated).

[16] Ascham, *The Scholemaster*, E2v (10).

uses to viscerally link the infant's sensory impressionability both to their moral "innocencie" and to their physical constitution.[17]

Supporters of experimental philosophy rejected many aspects of humoural theory, faculty psychology, and humanist pedagogy alike, but they continued to uphold the enhanced clarity and receptivity of children's sensory and cognitive processes, adding an emphasis on the retentiveness of their memories. For Cuff—again following Aristotle—one consequence of the sensory susceptibility characteristic of young children is that they also have "*slippery and short memories*", the reason being "their braines too great humidity, whereby it is disabled to keepe the impressions of the outward senses objects".[18] The humoural moistness of youth may have enhanced receptivity, but it diminished retention. Bacon, however, begged to differ. As he wrote in the *Novum Organum*, "things very firmly stamped upon a mind clear and untroubled both before and after—like things learnt in childhood … stick in the memory better".[19] Comenius reiterated the point, responding in 1650 to a critic who thought that his educational programme expected too much engagement with complex "technological, medical, optical, geometrical, astronomical, economic, and physical things" by insisting that, contrariwise, "childhood is the best moment for this kind of engagement, because the senses are so fresh that the objects go in more easily and adhere like glue to the memory".[20]

These qualities meant that children might have a privileged role to play in the practice of natural observation and experiment. We can see this in an account published in the *Philosophical Transactions* in 1676, in which the naturalist John Beale records an example of a phenomenon which baffled and fascinated many of his peers in the Royal Society: the emission of light by decaying meat and fish (a phenomenon which we know now arises from bioluminescent bacteria on the surface of the flesh).[21] In the

---

[17] Aristotle, *De anima* [*On the Soul*], 424a17–19.

[18] Cuff, *The Differences*, I7r (125). On children's mnemonic incapacity according to Aristotle, see Stephen Kidd: "since children are in flux and excessively moist, the thought-image (phantasma) underlying a memory cannot hold" (*Play and Aesthetics*, , 40; discussing Aristotle, *On Memory and Recollection*, 450a32-450b2). Erica Fudge discusses how children's supposed lack of memory is something they share with beasts in "Learning to Laugh", 279.

[19] Bacon, *Novum Organum*, 11.288.

[20] Comenius and Colbovius, *Sendschreiben*, 140. Quoted in Whitmer, "Reimagining the 'Nature of Children'", 121.

[21] On the early Royal Society's interest in various forms of luminescence, see Golinski, "A Noble Spectacle"; Daston, "The Cold Light of Facts"; and Swann, "From Philosopher's Stone to Phosphorus".

letter, Beale explains how "a fat Pork was killed for my Family", and the feet and chitterlings were boiled and pickled. Four days later, "all those parts of the guts, and the claws of the feet, which floated on the top of the pickle, began to shine ... as bright as the brightest Moon-shine". The parts of the pork "immersed under water", by contrast, "gave no light". Having "caus'd a Maid-servant to rub one of her hands upon the shining part [of the meat]", thereby transferring its luminosity, Beale describes how her hand continues to shine as she moves between differently lit rooms in his home. Subsequently, he attempts to establish whether a difference in temperature exists between the shining parts of the pork, and those submerged parts which remained unilluminated:

> Then I desired all the company, (whereof some were young children, which have the tenderest touch) to try, whether the most flaming parts had any perceptible degree of tepidity; all agreed, that they could feel no warmth. But I continued to direct them all to compare the dark parts with the most luminous, by that part of their fore-fingers, which hath the most tender perception, after 3 or 4 trials, all agreed still, that all parts of the Pork were manifestly gelid [icy cold]; but some thought, they perceived the luminous parts less gelid than the dark parts, others denied it: for my own part, I found not so much difference, as could clear me from suspecting a prepondering fancy.[22]

Beale's investigations into this strange occurrence take place in a distinctly collaborative, domestic context, with a maid-servant and "young children" as active participants.[23] Whereas, however, we might be tempted to suspect that Beale recruits the maid's assistance because he is initially reluctant to touch the luminous meat himself (its effects, after all, are still unknown), the subsequent involvement of the children (stage-managed by Beale, as he directs the company to use their "fore-fingers") has a clear epistemic benefit: having the "tenderest touch", they also have a keener sense of the meat's temperature. Beale's own "prepondering fancy", too—with its tendency to skew and render uncertain sensory evidence—is implicitly compared unfavourably to the children's lack of investment in or expectation about the "trials", or experiments, they participate in.

---

[22] Beale, "Two Instances of Something Remarkable", 601.
[23] Beale does not specify the identity of the children in question, but as he had at least three daughters and two sons, it is possible that they were his own offspring. See Woodland, "Beale, John".

Visual, rather than tactile, acuity is at stake in the naturalist John Ray's monumental, physico-theological *The Wisdom of God Manifested in the Works of the Creation* (1691), where the child's sensory acuity makes them not an active participant in the scene of experiment, but rather a paradigmatic experimental subject. Here, Ray expatiates on the perfectly designed physiology of the human eye, describing the pupil's response to light. Following the Jesuit physicist and astronomer Christophe Scheiner's *Oculus* (1619) and René Descartes' *La Dioptrique* (1637), Ray describes how the "*Iris of* the Eye hath a musculous Power, and can dilate and contract that round hole in it, called the Pupil. ... It contracts it for the excluding superfluous Light ... and again dilate[s] it for the apprehending Objects more remote, or placed in a fainter light".[24] As experimental verification of this process, he suggests that the reader:

> may, following *Scheiner* and *Des Cartes* their directions, take a Child, and setting a Candle before him bid him look upon it: And he shall observe his Pupil to contract itself very much, to exclude the light, with the Brightness whereof it would otherwise be dazled and offended ... Let the Candle be withdrawn, or removed aside, he shall observe the Childs pupil by degrees to dilate itself.[25]

Significantly, the experiments described by Scheiner and Descartes take as their subjects the eye not of "a Child", but rather of a man, *hominis*.[26] Ray, then, silently emends the original experiment to specify a subject whose eyes, undimmed by time, are most likely to respond as predicted, making the experiment a success.

We can also witness an emphasis on the sensory—and also, in this case, mnemonic—superiority of childhood in the reminiscences of the distinguished eighteenth-century Swedish botanist and physician Carl Linnaeus, in the autobiography he penned in his native language around 1770 (translated into English in 1798). Noting that "the same thing that is said of a poet, '*Nascitur non fit*, [born not made]' may be said without impropriety of our botanist", Linnaeus—writing in the third person—recalls how "from the very time that he first left his cradle, he almost lived in his

---

[24] Ray, *The Wisdom of God Manifested*, M7r (173). Scheiner and Descartes, of course, were not the first to notice the contraction and dilation of the iris around the pupil in response to changes in light: this had been recognized as far back as Galen (129–216 AD).

[25] Ray, *The Wisdom of God*, M7r-v (173–4).

[26] See Scheiner, *Oculus*, D4v (30).

father's garden … [and] never ceased harassing his father with questions about the name, qualities, and nature of every plant he met with".[27] By this time, in England, the trope of the precious boy-wonder was well established in scientific biographies: Richard Waller's "Life of Dr. Robert Hooke", included in Hooke's *Posthumous Works* (1705), records that "Indications of a Mechanick Genius appeared in him when very young", describing how, after his formal education was interrupted by ill health, Hooke "spent his time in making little mechanical Toys". John Ward's *The Lives of the Professors of Gresham College* (1740), meanwhile, reports that Christopher Wren made "remarkable" advancements in mathematical knowledge "before he was sixteen years old".[28] Linnaeus' memoir, however, stands out in describing his boyish aptitude for the natural sciences (rather than the mechanical or mathematical sciences) not as *in spite* of his youthfulness, but rather *because* of that youthfulness. "All the child's powers", professes Linnaeus, "both of mind and body, conspired to make him an excellent natural historian;—besides his retentiveness of memory, he had an astonishing quickness of sight". Linnaeus, that is, portrays his childhood self not as Waller and Ward portray Hooke and Wren, as a prodigy—unusually talented for his age—but simply as "an excellent natural historian" whose aptitude for botany was derived from a combination of an inborn interest with the child's excellent memory and sharp eyesight.[29] Indeed, in later life Linnaeus benefited from the heightened perceptions of youth in another way, organizing botanical expeditions with his students—boys who may have ranged from preteens to young men—which had both a pedagogical and a research purpose: on the expeditions,

[27] Linnaeus, *Diary*, 512. On the origins of this aphorism, see Ringler, *"Poeta Nascitur Non Fit"*.

[28] Waller, "The Life of Dr. Robert Hooke", in Hooke, *The Posthumous Works*, b1v (ii); Ward, *The Lives*, Bb2v (96). Ward also records that when William Petty was "was very young, he took great delight in conversing with artificers, and imitating their several trades, which he performed very dexterously at twelve years of age". *The Lives*, Kkk1r (217). On John Wilkins' nurturing of the youthful precocity of figures including Hooke and Wren, see Jardine, "The 2003 Wilkins Lecture"; on the precocity of Hooke and some of his scientific contemporaries more generally, see Jardine, *The Curious Life*, 91.

[29] Linnaeus, *Diary*, 513. In his biography of Linnaeus, Gunnar Broberg notes that "the image of Linnaeus as a child is an integral part of the cult of the man. As the 'child of nature', he gained by instruction in natural learning of a kind approved by Rousseau. … Even as an adult, he was thought naïve, innocent, and, hence, charming". Borberg, *The Man Who Organized Nature*, 29. However, on Linnaeus' antipathy to nature's playfulness, see Findlen, "Ludic Postscript", 62–3.

the students were encouraged to make their own observations, which "contributed significantly to [Linnaeus'] collections and his work".[30]

The attitudes of proto-empiricists such as Comenius, Sprat, Bacon, Beale, and Linnaeus were of course not universally shared, and the classical inheritance evident in Wright's *Passions of the Mind*, with its insistence on children's cognitive deficiencies and its depreciation of the senses, informed the thought of seventeenth-century rationalists—notably that of Descartes and his follower Nicolas Malebranche—in a very different way. Like Bacon and Comenius, Descartes and Malebranche avowed the extraordinary receptivity of children's senses, attributing these features to their humoural constitution. Thus, Descartes comments in a 1648 letter to fellow philosopher Antoine Arnauld that "as long as the mind is united to the body, it cannot withdraw itself from the senses". This is the case, for instance, when "it is attached to a brain that is too soft or damp, as in children".[31] As Descartes' censorious "too" implies, however, rationalist philosophers differed from their empiricist counterparts in their assessment of the implications of these characteristics, seeing them as an epistemological drawback, rather than as a benefit.

As Lisa Walters has shown, Descartes adheres to a broadly Platonic notion of infant cognition.[32] In Plato's *Meno* (c. 385 B.C.E.), the eponymous interlocutor challenges Socrates to prove that "what we call learning is recollection" (specifically, recollection of knowledge accrued in the soul's past lives, as the philosopher has argued).[33] In response, Socrates instructs Meno to summon an enslaved boy—a child with no prior mathematical training—and asks him to solve a basic geometry problem. Using his characteristic method of gentle interrogation to guide the boy, Socrates claims that his correct answer confirms that the child has recovered or recollected "the knowledge out of himself". His own questioning, he maintains, merely "stirred up" the knowledge the boy already possessed, but had forgotten. Although, in Platonic thought, this intrinsic wisdom is swiftly buried beneath the throng of sensations and passions which follow hard on the heels of embodiment, Meno must agree that—however

---

[30] Hodacs, "In the Field".

[31] Descartes to Antoine Arnauld, 29 July 1648, in *The Philosophical Writings of Descartes*, 3.356.

[32] Walters, "The Philosophy and Literature of Childhood Cognition", 204–5.

[33] Plato, *Meno*, 81C-E. On this episode, see Scott, *Recollection and Experience*, 33–38.

submerged—"the truth of all things that are is always in our soul", from infancy onwards.[34]

Similarly, according to Descartes, writing in a 1641 letter to the anonymous critic known as Hyperaspistes, "an infant" (a term which here includes the foetus in its mother's womb, as well as the newborn) "has in itself the ideas of God, of itself and of all such truths as are called self-evident" (with the important qualification that for Descartes those truths are implanted by God himself, rather than being—as in *Meno*—a residue of the soul's past lives).[35] Just as for Plato, too, the infant's ability to attend to these ideas is frustrated by its entrapment in what Descartes calls "the prison of the body", for "a mind newly united to an infant's body is wholly occupied in perceiving in a confused way or feeling the ideas of pain, pleasure, heat, cold and other such ideas that arise from its union and, as it were, intermingling with the body".[36] The problem is compounded in childhood, when sensory errors gradually calcify into mistaken beliefs—a theory which Descartes elaborates most fully in his *Principles of Philosophy*, first published in Latin in 1644. "In our childhood", writes Descartes, "the mind was so immersed in the body that it perceived many things vividly but nothing clearly".[37] For example:

> because the light coming from the stars appeared no brighter than that produced by the meager glow of an oil lamp, it [i.e. the child's mind] did not imagine any star as being any bigger than this. And because it did not observe that the earth turns on its axis or that its surface is curved to form a globe, it was rather inclined to suppose that the earth was immobile and its surface flat. Right from infancy our mind was swamped with a thousand such preconceived opinions; and in later childhood, forgetting that they

---

[34] Plato, *Meno*, 82A–86B.

[35] Descartes to Hyperaspistes, August 1641, in *The Philosophical Writings of Descartes*, 3.190. On Hyperaspistes, see Smith, "Hyperaspistes", 384–86.

[36] Descartes to Hyperaspistes, August 1641, in *The Philosophical Writings of Descartes*, 3.190. On Descartes' ideas about foetal consciousness and childhood error, see Wilkin, "Descartes, Individualism", 104–09, and Harrison, *Coming To*, 17–20.

[37] Descartes, *Principles of Philosophy*, in the version presented by Jonathan Bennett at www.earlymoderntexts.com. Bennett's digital transcription is a modification of the translation by Cottingham *et al.* in Descartes, *The Philosophical Writings of Descartes* 1.208. It is preferred here on the basis of Bennett's more cogent translation of Descartes' Latin adjectives *clarus* and *distinctus*, inaccurately translated by Cottingham *et al.* as "clearly" and "distinctly" (presumably, as Bennett speculates, on the basis of "the physical similarity of the words").

were adopted without sufficient examination, it regarded them as known by the senses or implanted by nature, and accepted them as utterly true and evident.[38]

In this context, youthful sensitivity is an obstacle to knowledge of the natural world, not an advantage, as the confused and misleading bodily sensations which "swamp" the child obscure their inborn grasp of rational "truths". As Malebranche puts it in his *Search After Truth* (1674–1675), translated into English by Thomas Taylor:

> Children don't appear very proper for the Meditation of Truth, and for abstracted and elevated Sciences, because the Fibres of their Brain, being very delicate, they are easily agitated, even by the weakest and least sensible Objects; and their Soul necessarily having Sensations proportionated to the agitation of these Fibres, she lets go her Metaphysical Thoughts, and pure Intellections, to apply her self only to her own Sensations. Thus, it seems, Children cannot consider the pure Idea's of Truth with sufficient attention, being so often and easily disturbed by the confused Idea's of their Senses.[39]

For empiricists such as Comenius, Bacon, and some members of the early Royal Society, I have suggested, childish sensitivity was to some degree exemplary, offering a model of impressionability and retentiveness to which the natural philosopher could only aspire. By contrast, for Descartes and Malebranche, the child's heightened sensitivity to the material world is an epistemological disaster, distracting them from the "Metaphysical Thoughts, and pure Intellections" they are born with.

The distinction, however, is not quite as stark as it initially appears. For a start, Comenius—like Plato and Descartes—believed that children possess an intrinsic, inborn form of knowledge, differing from the French philosopher only in his optimism about their capacity to retain that

---

[38] Descartes, *Principles of Philosophy*, in *The Philosophical Writings*, 1.219, trans. Cottingham *et al.* All future references to the *Principles* are to this edition. See also Descartes, *Meditations on the First Philosophy*, 65: "Although a star has no greater effect on my eye than the flame of a small light, that does not mean that there is any real or positive inclination in me to believe that the star is no bigger than the light; I have simply made this judgment since childhood onwards without any rational basis".

[39] Malebranche nonetheless acknowledges that "it is more easie for a Child of seven years to be deliver'd from the Errors whereinto the Senses lead it, than for a person of Sixty, who has all his life time followed the prejudices of Infancy". Malebranche, *Malbranche's Search after Truth*, M2r-v (163–4).

knowledge, and his conviction that the senses may assist, rather than sabotage, efforts to do so. Making reference to the *Meno* episode in response to criticism that his educational programme was too difficult, he comments that "the wise Socrates thought otherwise, asserting that also an eight-year-old boy is capable of answering all the questions of Philosophy … because the inner light of the understanding, when properly invoked or elicited, is available to everyone".[40]

Robert Boyle, moreover, tentatively affirms Descartes' ideas about childhood error in *Some Considerations About the Reconcileableness of Reason and Religion* (1675), offering in support of the suggestion that "the very Body of Mankind may be embued with Prejudices, and Errors … from their Childhood, and some also ev'n from their Birth" the example of "Monsieur *Des Cartes*", who:

> begins his Principles of Philosophy with taking notice, That, because we are born Children, we make divers unright Judgments of things, which afterwards are wont to continue with us all our Lives, and prove radicated Prejudices, that mislead our Judgments on so many occasions, that he elsewhere tells us, he found no other way to secure himself from their Influence, but once in his Life solemnly to doubt of the Truth of all that he had till then believ'd, in order to the re-examining of his former Dijudications.[41]

At the same time, conversely, Descartes' famous method also contains its own tensions and contradictions. As Boyle's account suggests, its first stage—radical doubt—is explicitly framed as a means to clear the philosopher's mental ground of the chaff of childish errors. As Descartes writes in the *Principia*, "the only way of freeing ourselves" from the "preconceived opinions" we accrue "as infants" is "to make the effort, once in the course of our life, to doubt everything which we find to contain even the smallest suspicion of uncertainty".[42] Similarly, "a philosopher … should never rely on the senses, that is, on the ill-considered judgements of his childhood,

---

[40] Colbovius and Comenius, *Sendschreiben*, 140; quoted in Whitmer, "Reimagining the 'Nature of Children'", 121.

[41] Boyle, *Some Considerations about the Reconcileableness*, C6v-7r (28–9).

[42] Descartes, *Principles of Philosophy*, in *The Philosophical Writings*, 1.193. See also his 1638 letter to Henri Regnier, intended for Alphonse Pollot: "those who want to discover truth must above all distrust opinions rashly acquired in childhood". Descartes to Regnier, April/May 1638, in *The Philosophical Writings*, 3.99.

in preference to his mature powers of reason".[43] Here, exercising doubt is a means of achieving maturity: scepticism liberates the philosopher from childish preconceptions. Nonetheless, at the same time there is a sense in which stripping the mind of its accretions of sensory error is also a return to infancy in a purified form: an attempt to recover the clarity of inborn reason before it became clouded by the passage of time. In the words of Gareth B. Matthews, "Descartes taught us to do philosophy by 'starting over' ... I am to make a fresh beginning ... In a certain way, then, adult philosophers who follow Descartes in trying to 'start over' are trying to make themselves as little children again, even if only temporarily".[44] For Descartes, the child's enmeshment in the flesh is a terrible static, drowning out and distorting the promptings of reason. And yet, embedded in his own method is a yearning to return to a primordial point in the philosopher's own development, to a moment even prior to birth: the split second, perhaps, before the body "swamps" the souls with sensations, and the infant's perception of "the truths that are called self-evident" remains pristine.

*   *   *

In the seventeenth century, I have suggested, an association emerged in the work of figures associated with the early Royal Society between children's instinctive use of their senses as a means of understanding the world, and the Society's own preference for immediate sensory experience over the textual authority of the ancients or unfounded speculative reasoning as a source of knowledge. In this context, actual children's senses were also thought to have epistemic advantages over those of adults, having a clarity and keenness lost with the degradations of age.

In the later eighteenth and nineteenth centuries, the notion that there is something special about the purity and receptivity of the child's senses

---

[43] Descartes, *Principles of Philosophy*, in *The Philosophical Writings*, 1.221–22.

[44] Matthews, *The Philosophy of Childhood*, 18. Matthews identifies as the target of Descartes' doubt "the correctness of what my teachers have taught me, or what the society around me seems to accept"; in fact, the target is rather the errors the philosopher's own mind has built from the blocks of the fallible senses. Walter Kohan suggests that Socrates took "a childlike position in relation to knowledge", arguing that "for Socrates the truest relationship to knowledge—knowing nothing and not believing he knows—is childlike, playful", in Kohan, "Childhood, Philosophy, and the Polis", 35.

reached its apogee in the Romantic idealization of childhood.[45] Towards the end of the seventeenth century, the "sensualist" psychological model which originated with Aristotle, according to which the infant's mind is a blank slate awaiting sensory impression, received definitive new expression in the work of John Locke. "Let us suppose the mind", Locke invites his readers in his *Essay Concerning Human Understanding* (1693), "to be, as we say, white paper, void of all characters", awaiting the informing impressions of "Experience".[46] In the following century, Jean-Jacques Rousseau, too, blended sensualist psychology and Jesuit pedagogy with Baconian empiricism, instructing the would-be teacher in his *Émile, or On Education* (1762), that "in the first operations of the mind", the senses should always be the child's "guides". "Purely speculative knowledge", Rousseau advises, in words which might have drawn a nod of agreement from Comenius, "is hardly suitable for children. … In the quest for the laws of nature, always begin with the phenomena most common and most accessible to the senses".[47] Whereas, however, some earlier thinkers had suggested that children's sensory impressionability meant they had a role in the *production* of new knowledge about nature, Locke and Rousseau's interest was firmly in the ways in which children *acquire* understanding of the world.

What, then, became of the idea that children not only learn through their senses, but also have something to teach their elders as they do so? To track its endurance, we may turn to a man roughly contemporaneous with Rousseau: the Scottish Enlightenment philosopher Thomas Reid.

---

[45] On the "idealized, nostalgic, sentimental figure of childhood" in Romantic literature and culture, "characterized by innocence, imagination, nature and primitivism", see Rowland, *Romanticism and Childhood*, 9.

[46] Locke, *An Essay*, Book II, Chapter I, Section 2. See also *Some Thoughts Concerning Education* (1693), where Locke considers the child "as white paper, or wax, to be moulded and fashioned as one pleases", recommending that teaching should begin as much as possible "with that which lies most obvious to the senses; such as is the knowledge of minerals, plants, and animals, and particularly timber and fruit-trees … where a great deal may be taught a child, which will not be useless to the man". Locke, *Some Thoughts Concerning Education*, 179, 138. On the process by which children acquire knowledge through sensation, and on the importance of neonatal experience in particular to Locke's epistemology, see Harrison, *Coming To*, Chapter 5.

[47] Rousseau, *Émile*, 168, 177. Rousseau is, however, highly critical of experimentalism's reliance on sophisticated tools and sensory prosthetics: "All the instruments invented to guide us in our experiments and to take the place of accuracy of the senses cause the senses to be neglected. … The more ingenious our tools, the cruder and more maladroit our organs become". *Émile*, 176.

Reid's "common sense" philosophy—also heavily influenced by Bacon—rejected the scepticisms of Descartes and David Hume, developing instead a realist theory of perception according to which we can trust our senses, along with our innate faculty of judgement, to give us accurate knowledge about the world.[48] "The faculties of the human mind", claims Reid in *An Inquiry into the Human Mind on the Principles of Common Sense* (1764), "as they are naturally endowed, have an innate power of perceiving and understanding certain fundamental truths"; he therefore resolves to "take my own existence, and the existence of other things, upon trust; and to believe that snow is cold, and honey sweet".[49] This does not mean, of course, that error is impossible, but rather that it is an aberration from our "natural" state: a product of over-sophistication, rather than ignorance. "It is genius, and not the want of it", Reid declares, "that adulterates philosophy, and fills it with error and false theory". Consequently, the quest for truth is also a process of regression: "A man that is grown up in all the prejudices of education, fashion, and philosophy [must] … unravel his notions and opinions, til he finds out the simple and original principles of his constitution": principles which are, for Reid, embodied in the child.[50]

We can see this in Reid's discussion of sensory error, which, he submits, often arises not from the senses themselves, but rather from misguided interpretations of sensation—especially, a tendency to leap too precipitously from perceptions *per se*, to speculative assumptions about the things prompting those perceptions. For instance, Reid writes of the "sensation of hardness" that one gets "by pressing one's hand against the table" that "it is one thing to have the sensation, and another to attend to it and make it a distinct object of reflection. … There is no sensation more distinct, or more frequent; yet it is never attended to, but passes through the mind instantaneously, and serves only to introduce that quality of bodies which … it suggests".[51] These "habits of inattention", Reid explains, are "acquired very early [in our lives]". They are not, however, as fully engrained in the young as in the old, as "the novelty of this sensation will

---

[48] For the influence of Bacon (and Isaac Newton) on Reid, see Laudan, "Thomas Reid and the Newtonian Turn", and Wood, "Thomas Reid and the Culture of Science", 68–9. On Reid's rejection of scepticism, see De Bary, *Thomas Reid and Scepticism*.

[49] Reid, *An Inquiry into the Human Mind*, 26.

[50] Reid, *An Inquiry into the Human Mind*, 17.

[51] Reid, *An Inquiry into the Human Mind*, 56–7.

procure some attention to it in children at first". As such, "we must become as little children again, if we will be philosophers".[52]

It is not only in their attentive lingering with sensations that the common-sense philosopher must emulate children; moreover: he must also imitate their predisposition to trust. "The common sense of mankind", Reid declares, "is to be trusted when it tells us that our senses do not deceive us without a cause". If he is to overcome the ersatz allure of scepticism, then, the common-sense philosopher must re-assume the child's credulity: just as, as a child, he "believed by instinct" whatever he was told by "my parents and tutors", so too does the adult Reid "deal with the Author of my being", "yield[ing] to the direction of my senses, not from instinct only, but from confidence and trust in a faithful and beneficent Monitor, grounded upon the experience of his paternal care and goodness".[53]

For Reid, then, children are models of both sensory and epistemological probity. What, though, of morality? Earlier in this chapter, I commented that for humanist educators such as Erasmus and Ascham, children's putative openness to pure sensation was contiguous with their "innocencie". In their time-limited exemption from what Erasmus calls the "tares and vices" of sin, children perceived more astutely and learned more quickly than their adult counterparts. In the next section, I explore how, across the course of the seventeenth century, poets and experimentalists increasingly identified the innocence of children not only as the source of their intellectual pliability, but as akin to a kind of epistemological purity akin to what we now know as "objectivity"—a freedom from the forms of "prepondering fancy" that led John Beale to distrust his own perception of temperature as he dabbled in a radiant pot of pickled pork.

---

[52] Reid, *An Inquiry into the Human Mind*, 58.
[53] Reid, *An Inquiry into the Human Mind*, 173.

CHAPTER 3

Virgin Minds: Innocence and Objectivity

Abstract This chapter explores connections between ideas about childish innocence and the emergence in the seventeenth century of something like the modern scientific virtue of objectivity, as well as the implications of this link for our understanding of the historical development of the latter concept. Defined partly as freedom from worldly prejudice and preconception, as well as a state of prelapsarian moral perfection, childlike innocence was increasingly held up as an epistemological as well as an ethical ideal, facilitating perceptual lucidity and accuracy. In particular, childish ignorance of worldly standards of value was both used to satirize the similarly indiscriminate collections of seventeenth-century philosophers and as evidence for their innate insight and piety in preferring divinely created things over the bogus glitter of earthly treasures. Ancient ideas about the autotelic character of play were also important in this regard, allowing thinkers to postulate a form of experiment which was productive precisely insofar as it was purposeless.

Keywords Children • Science • Natural history • Experiment • Royal Society • Innocence • Objectivity • Thomas Traherne • Robert Boyle • Walter Charleton • Virginity • Cultures of collecting • Occult • Iconoclasm • Play • Francis Bacon • Judith Drake • Robert Hooke

© The Author(s), under exclusive license to Springer Nature Switzerland AG 2024
E. L. Swann, *Science as Child's Play in Seventeenth-Century England*, https://doi.org/10.1007/978-3-031-75849-2_3

In the introduction to this book, I commented on Francis Bacon's figuration, in the Preface to the *Instauratio Magna*, of natural philosophy as "an innocent friendly children's game of hide and seek" played with none other than God himself. Writing in a period when curiosity was viewed with profound suspicion as a sinful outgrowth of Adam and Eve's disastrous desire for "knowledge of good and evil", Bacon responds by reframing the pursuit of natural knowledge as innocuous as child's play: a quest for a God who, like any other obliging father, has hidden himself for fun. This chapter contends that Bacon's foregrounding of childish innocence is not only a part of his promotional strategy for natural philosophy: it also reflects a deep investment in the epistemological possibilities of youthful purity. In a range of devotional, literary, and natural philosophical works in the latter half of the seventeenth century, I propose, the simplicity of the child comes to represent a disposition which is meritorious not only in its moral excellence but also in its combining of curiosity and inexperience, enabling a revelatory perspective unskewed by worldly conventions and preconceptions.

Exemplary here is the Anglican cleric and late metaphysical poet Thomas Traherne, whose work frequently depicts what I suggested in the previous chapter, Descartes craved: the primitive purity of prenatal existence.[1] Specifically, Traherne describes, with near-hallucinatory intensity, not only his childhood—which he represents, in stark defiance of the Church of England's adherence to the doctrine of original sin, as an idealized state of prelapsarian innocence—but also his experiences of infancy (including gestation in the womb), which he professes to remember by the "Special favor" of God.[2] In his poem "The Preparative", Traherne describes embryonic experience prior to the infant's conscious awareness of its own embodiment, "before I knew my Hands were mine", in terms which syncretize the mystical, Neoplatonic model of innate ideas with the

---

[1] Descartes exerted a significant influence on Traherne; see Harrison, *Coming To*, 126–30; Watson, *Back to Nature*, 300–301, and Johnston, "Heavenly Perspectives, Mirrors of Eternity", 393 (which focuses on Traherne's adaptation of Descartes' optics). On the child in Traherne as "an emblem of privileged sensitivity and freshness of sensation", see Davis, "Brilliance of a Fire", 384–5.

[2] Traherne, *Centuries of Meditations* 3.1, in *The Works*, 5.93. On Traherne's lack of emphasis on original sin, see Wilmot, *Voluble Soul*, p.4. On Traherne's "claims to remember neonatal experience", see Harrison, *Coming To*, esp. 149–161.

Aristotelian model of the infant's mind as a blank slate.[3] Not yet "prepossest" with the "Dross" of the flesh, the foetus is nonetheless imbued with divine powers of visual perception: "then was my Soul", Traherne's speaker claims, "A Living Endless Ey", able to apprehend "the fair Ideas of all Things". In contrast to Descartes' suggestion that the infant is flooded with distracting sensations even in the womb, Traherne suggests that in this state of supernatural insight other sensory desires or passions have no purchase: "I then no Thirst nor Hunger did conceiv; /No dull Necessity". Unfettered from the "vain Affections" of "Earthy State", the infant soul is also, simultaneously, a *tabula rasa* consisting of:

> A Disentangled and a Naked Sence,
> A Mind thats unpossest,
> A Disengaged Breast,
> An Empty and a Quick Intelligence...

Insisting on the absolute perfection of the infant's understanding, Traherne blends divinely infused knowledge with the sensory receptivity of "a Naked Simple Pure *Intelligence*". Unspoiled by worldly prejudices or desires, the senses and mind of the unborn child are "Pure Empty Powers" which "nothing loath", and which subsequently reflect everything they encounter "like the fairest Glass", without obstruction or distortion.[4]

Timothy Harrison has argued persuasively that Traherne's obsession with the earliest stirrings of consciousness in the womb is propelled by his conviction that "since this is as close to God as living beings can come, our first thoughts are the best we can think", stressing in particular Traherne's efforts to recapture "a feeling of bliss and an admiration of the world into which he has just awakened" and his sense "that his unique memories of prenatal and infant experience" offer a source of self-knowledge as "the

---

[3] Thomas Traherne, "The Preparative", in *The Works*, 6.12–13. Traherne's interest in combining Neoplatonic innatism with Aristotelian sensualism is also evident in the *Centuries of Meditations*, where he makes claims both for inborn knowledge ("at my entrance into the World ... my knowledge was Divine: I knew by Intuition"), and for the *tabula rasa* model of cognition ("an Empty Book is like an Infant's Soul, in which any Thing may be written"). See Traherne, *Centuries* 3.1 and 1.1, in *The Works*, 5.93, 5.7. For an astute discussion of Traherne's "broadly Piconian synthesis of Platonic anamnesis and Aristotelian tabula rasa ontologies", see DeFries, "Love, Capacity, and Traherne's Idea of the Book", 104–5, 112–15; see also Harrison, *Coming To*, 157–60.

[4] Traherne, "The Preparative", in *The Works*, 6.12–13.

best way to gain access to one's true nature".[5] Traherne was keenly (though not uncritically) interested in contemporary experimental philosophy, however, and his interest in the unsullied purity and receptivity of the infant soul can also be understood in this context: not only as a source of self-knowledge and recaptured affective awe but also as reflecting a broader preoccupation amongst seventeenth-century natural philosophers with childish innocence as an epistemologically privileged state when it comes to acquiring knowledge about the natural world.[6]

In this regard, it is important Traherne often suggests both that a degree of prenatal lucidity persists into childhood and that it may be recaptured, partly at least, in adulthood. As he asserts in his series of short theological and biographical reflections, *Centuries of Meditations*, it is this state of being that Christ urged us to aspire to in the Gospel of Matthew, when he told the disciples that "Except ye be converted, and become as little children, ye shall not enter into the kingdom of heaven".[7] It is, Traherne continues:

> Not only in a careless reliance upon Divine Providence, that we are to becom Little Children, or in the feebleness and Shortness of our Anger and Simplicity of our Passions: but in the Peace and Purity of all our Soul. which Purity also is a Deeper Thing then is commonly apprehended … all our Thoughts must be Infant-like and Clear; the Powers of our Soul free from the Leven of this World, and disentangled from mens conceits and customs. Grit in the Ey or the yellow Jaundice will not let a Man see those Objects truly that are before it. And therfore it is requisit that we should be as very Strangers to the Thoughts Customs and Opinions of men in this World as if we were but little Children.[8]

Insisting that the meaning of this verse is "Deeper" than is usually appreciated, Traherne implicitly takes aim at no lesser exegetical authorities than Augustine and John Calvin, both of whom had offered readings of this passage. Particularly striking in the context of these previous

---

[5] Harrison, *Coming To*, 150–51.

[6] Scholarship addressing Traherne's (often critical) engagement with various aspects of the new philosophy includes, *inter alia*, Nicolson, *The Breaking of the Circle*; Balakier, "Thomas Traherne's Dobell Series"; Clucas, "Poetic Atomism"; Sawday, *The Body Emblazoned*, 261–66; Gorman, "Thomas Traherne and 'Feeling Inside the Atom'"; Partner, *Poetry and Vision*; Rimmer, *Greening the Children of God*; and Willmott, *The Voluble Soul*, Chapter 9.

[7] Matthew 18:3 (KJV).

[8] Traherne, *Centuries of Mediations* 3.1, in *The Works*, 5.96.

interpretations is Traherne's insistence on the excellence not only of the child's moral instincts but also of their cognitive faculties.

In the *Confessions*, Augustine claims that Christ bestows "the Kingdom of God" on children and the childlike not because of their "innocencie", but rather because God "hast allowed the Character of humility in the stature of Childehood"—a position consistent with his ideas about the engrained corruption even of very young children.[9] For Augustine, there is nothing meritorious about childhood, and Christ merely evokes it in order to express the requirement for meekness and modesty. Calvin's reading is similar but distinct. Glossing the passage by reference to St Paul's words to the Corinthians—"Brethren, be not children in understanding: howbeit in malice be ye children, but in understanding be men"—he proposes that we should not aspire "to be children in understanding, but in malice … because there doth yet raigne so great simplicitie in infants, that they knowe not the degrees of honours nor the swellinges of pride".[10] Although he stops short of attributing "innocencie" to children, Calvin is nonetheless more positive than Augustine about their affective disposition, alluding to a lack of "malice". At the same time, he also places a heavy emphasis on humility ("they knowe not … the swellinges of pride"), and follows Paul in dismissing the idea that there is anything laudable about their "understanding".

Traherne, by contrast, explicitly counters the Pauline suggestion that it is not the child's "understanding" but only their affective "simplicitie", their lack of cruelty and pride, that is admirable. "It is not only in … [the] simplicity of our passions" that we should emulate children, he maintains, but in the purity of our perceptions and judgements: "all our thoughts must be infant-like and clear". Here, a return to the cognitive purity of childhood is a prerequisite both of spiritual salvation and of accurate perception in this world: Traherne's iteration of the commonplace that "grit in the eye or yellow jaundice will not let a man see those objects truly that are before it" implies that the adult "thoughts, customs, and opinions" which grit and jaundice analogize are a barrier to seeing physical things, as well as divine.

---

[9] Augustine, *Confessions*, D6v (60). For a discussion of this passage, see Willmott, *Voluble Soul*, 42–3.

[10] Calvin, *A Harmonie Upon the Three Evangelists*, Hh3r. Calvin cites 1 Corinthians 14:20 (KJV). On this passage, see Pitkin, "'The Heritage of the Lord'", 65.

Traherne's interest in the epistemological possibilities of naïve or "unprejudiced" perception was shared by advocates and practitioners of the mechanical and experimental philosophies. The physician and natural philosopher (and later, Fellow of the Royal Society) Walter Charleton, for example, proclaims in his 1654 outline of Epicurean atomism, *Physiologia* (a simplified and augmented translation of Pierre Gassendi's *Animadversions* of 1649) that there is "no man so fit to receive and retain the impressions of *Truth*, as He, who hath his Virgin mind totally dispossessed of *Praejudice*".[11] Similarly,

> Some *Texts* there are in the *Book of Nature*, that are best interpreted by the sense of the *Vulgar*, and become so much the more aenigmatical, by how much the more they are commented upon by the subtile discourses of the *Schools*: their over-curious *Descants* frequently rendring that *Notion* ambiguous, complex and difficult, which accepted in its own genuine *simplicity*, stands fair and open to the discernment of the unpraejudicate.[12]

Although Charleton makes no explicit mention of children, his claims about the advantages of "Virgin" and "unpraejudicate" minds when it comes to understanding nature's book endorse naïve experience in a way which has the potential to evoke the "simplicity" of the childish—or child-like—observer, as well as that of the putatively "vulgar" tradespeople, artisans, and agricultural workers Charleton and his fellows claimed to welcome into the experimental fold.

Charleton primarily uses the word "Virgin" in a figurative sense, to indicate a mind pure of preconceptions. Whilst in this period "innocence" was not as closely linked to sexual inexperience as it is today, however, physical virginity or chastity was nonetheless an aspect of moral virtue, and as such was valued by early modern virtuosi.[13] This is most famously the case regarding the much-vaunted virginity of another figure influenced by Gassendi's reformulation of ancient atomism on Christian lines: Robert

---

[11] Charleton, *Physiologia*, O2r (99).

[12] Charleton, *Physiologia*, K4r (72).

[13] Despite the fact that, as Joanna Picciotto observes, the "virtuosi eroticized natural philosophy", they also celebrated "the conquered and chastened sensitive body"; "the quintessential Christian virtuoso seeks to stamp out all thought of the carnal body of private man". Picciotto, *Labours of Innocence*, 230–31. On the perceived importance of physical abstemiousness or even asceticism for early modern experimental philosophy (with a focus on dietetics, rather than sex), see Shapin, "The Philosopher and the Chicken".

Boyle. In his funeral sermon for Boyle, Boyle's friend and spiritual advisor Gilbert Burnet drew on Boyle's own recollections, in his autobiographical *An Account of Philaretus in his Minority* (written c.1648–1649), of how in his youth he visited the bordellos of Italy without any impulse to succumb to the carnal allurements they offered. For Burnet, this sensory self-mastery is evidence of Boyle's preternatural maturity. Whilst "many [men] are Children to their Lives end", Burnet proclaims, "a good man is one that ... renders himself as clean and innocent ... as he can possibly make himself to be ... [and] that rises as much as he can above his body, and above this world, above his senses".[14] As such a man, Boyle "passed through the Youthful parts of life, with so little of the *Youth* in him that in his travels while he was very young and wholly the Master of himself he seemed to be out of the reach of the disorders of that Age, and those Countries through which he passed".[15] Gilbert associates Boyle's lifelong virginity not with childish inexperience, but rather with moral seniority: in resisting the fleshpots of Venice and Rome, Boyle transcended the passionate appetites of sinful youth. Elsewhere, however, we find Boyle battling a different (and more potent) form of temptation: namely, the desire for occult knowledge. And in this context, Boyle's chastity functions slightly differently, both as an index of his moral superiority and as conferring a tantalizing—if dangerous—form of supernatural insight usually only available to children.

In the document known as the "Burnet Memorandum"—a manuscript record of Burnet's notes on his biographical interviews with Boyle—Burnet relates a story originally told to Boyle by a "very vertuous" gentleman "of the Royall society". Travelling in Venice, the gentleman (who remains anonymous in Burnet's account) took with him "a locket of Diamond" containing hair gathered from the combs of "a lady" with whom he was "much in love". Finding that the locket had been stolen, the gentleman resorted to supernatural aid to find the culprit, seeking out "a Priest who ... had a Magicall glasse" which would relinquish its secrets only to "Virgins". The gentleman therefore procured the help of "a young girle ... about nine year old", who discovered the thief (a serving boy) in the mirror, convincing the gentleman of the authenticity of the vision by describing "the locket and the colour of the hair and also a picture of the

---

[14] Burnet, *A Sermon*, A3v, B1r-v (6, 9–10).
[15] Burnet, *A Sermon*, C4r (23).

ladies losst at the same time with some other things one of them she could not well describe for she had never seen any of the sort".[16]

Aspects of this strange, fairytale-like fable are conventional within the framework of Renaissance occultism. In particular, the mirror's refusal to yield its secrets to anyone but a virgin reflects an established belief that, as Cornelius Agrippa writes in his *Three Books of Occult Philosophy*, first published in full in 1533 and translated from Latin into English in 1651, "innocency of the mind" as well as physical "purity, [and] chastity" was necessary for anyone who hoped to communicate with "divine spirits".[17] In this context, stories about scrying mirrors which would only reveal their secrets to children were not uncommon: in the late sixteenth century, for instance, the mathematician, astronomer, and occultist John Dee recounted a story about "a Gentleman of *Norimberg*" who "had a Crystal" with the ability to reveal "any thing past or future", but only to "a young Boy (*Castum* [i.e. chaste], one that was not yet of Age, *&c.*). ... No other but a Boy, so qualified, could see any thing in it".[18] Indeed, in April 1587, Dee enlisted the help of his own seven-year-old son, Arthur, in "the service of Seeing and Skrying from God".[19] The idea that the child's moral purity—particularly their virginity—conferred privileged insight when it came to the supernatural realm was, then, well-established.

In the story related by Boyle via Burnet, the young girl's success in identifying the source of the locket's loss leads the gentleman to engage her services in the search for other forms of knowledge: subsequently, "he came often with his Interpreter the girle who was to see for him and among things having got a processe of the Philosophers stone he brought it with him and made the Priest ask if it was a good one". This effort to leverage the child's visions for alchemical ends is ultimately thwarted: the girl reports that the spirits she sees in the mirror—who themselves take the form of "pretty boies"—became "very angry" at the question, and "so the Priest said he must presse it no further least the spirits should tear them to

---

[16] Burnet, "The Burnet Memorandum", 31–32. On this episode, see Hunter, "Alchemy, Magic and Moralism", 390–92, which speculates that the gentleman in question might be Boyle's friend, Sir Robert Southwell.

[17] Agrippa adds, furthermore, that possession of "a pure minde" as a precondition for communion with good spirits explains their tendency to appear to "children, women, and poor and mean men". Agrippa, *Three Books of Occult Philosophy*, F7v (78) and Gg1v (450).

[18] Dee, *A True & Faithful Relation*, G1v (unpaginated).

[19] Dee, *A True & Faithful Relation*, *A222v (unpaginated).

pieces".[20] The crux of Boyle's tale, however, is not the gentleman's failure in this regard, but his own resistance to the use of supernatural means to pursue potentially transgressive natural knowledge: an option which is nonetheless available to him precisely because of his own childlike sexual innocence. As Burnet reports at the end of the anecdote:

> Another Gentleman not only told him [Boyle] stories like this but brought him a glasse in which he did not doubt but he was a Virgin and so could satisfy himselfe … he had the greatest Curiosity he ever felt in his life tempting him to look into it and said if a Crown had been at his feet it could not have wrought so much on him[:] but he overcame himselfe which he accounted the greatest Victory he had ever over him selfe.[21]

Boyle's moral probity makes it possible for him to use the scrying "glasse", as his virginity preserves in him the childlike purity which is a precondition of communion with spirits in the natural magic tradition. Simultaneously, that same piety prevents him from doing so, as—uncertain of its permissibility—he ultimately resists his fierce "Curiosity". Whereas Burnet described Boyle's virginity as a product of his mature self-restraint, then—a rejection of the undisciplined appetites of youth—Boyle also hints at an association with the innocence and perceptual clarity of children.

Boyle's ambivalence about youth is also evident in his own natural philosophical works—where, moreover, he also expresses a sense of his own enduring juvenility. Burnet's claim, in the funeral sermon, about Boyle's youthful maturity echoes the passage of the *Occasional Reflections* discussed in the introduction to this book, where Boyle, writing in his late teens or early twenties, is eager to distance himself from the child who cries because he cannot possess the stars, emphasizing his own manly self-control in the quest for astronomical knowledge. In the middle phase of his career, however, we find Boyle asserting quite the opposite: namely, his own persistent youthfulness. Thus, in the preface to his *Certain Physiological Essays* (1661), published in his thirty-fourth year, we find Boyle attributing any flaws in the work to his "being yet but very young, not only in Years, but, what is much worse, in Experience", and therefore deficient in "the expected Maturity of Age and Judgment".[22] Elsewhere, he presents his own putative inexperience more positively, not as

[20] Burnet, "The Burnet Memorandum", 31–32.
[21] Burnet, "The Burnet Memorandum", 32.
[22] Boyle, *Certain Physiological Essays*, B1v (2).

contiguous with a lack of judgement but rather as conferring the ability to resist the premature systematization of experimental findings to which members of the Royal Society, following Bacon, were so hostile. Suggesting, in his *Considerations Touching the Usefulness of Experimental Natural Philosophy* (1663), that "chymistry" has the capacity to improve the practice of physic, Boyle (writing in his thirty-seventh year) nonetheless avows that "I am much too young, too unlearned, and too unexperienced, to dare to be dogmaticall in a matter of so great moment".[23] Here, Boyle claims his own youthful inexperience—hence, humility—as an epistemic virtue in the practice of experimental science, a defence against the obstinate dogmatism so deplored by Bacon and his followers.

Indeed, the figure of the innocent child stands at the very centre of what is often taken as pivotal moment in the history of the development of scientific objectivity: Bacon's famous discussion of the four "idols of the mind" in his *Novum Organum*.[24] Immediately after his discussion of how these idols—which include human nature itself; the individual's subjective, irrational preferences and aversions; the corruptions of social and linguistic custom; and the influence of outmoded philosophical dogmas—have hitherto obstructed human progress in the sciences, Bacon concludes:

> The individual kinds of *Idols* and their trappings … must be forsworn and renounced with unwavering and solemn resolve, and the intellect must be thoroughly freed and purged of them, since entrance into the Kingdom of Man, which is founded on the sciences, differs little from that into the Kingdom of Heaven, *into which none enters except in the likeness of a little child.*[25]

As Joe Moshenska has recently argued, children quite literally played a role in Reformation iconoclasm: "Holy things", he demonstrates, "were made into playthings with sufficient frequency for it to be considered an established and recognized part of iconoclastic practice".[26] Here, in Bacon's religiously charged language, natural philosophy is well overdue a similar clear-out of idolatrous trash and children likewise have significant part in

---

[23] Boyle, *Some Considerations Touching the Usefulnesse*, Bb2r (203).

[24] On Bacon's discussion of the idols of the mind as a key moment in the development of the notion of objectivity, see Zagorin, "Francis Bacon's Concept of Objectivity". Zagorin challenges Lorraine Daston and Peter Galison, who argue that Bacon wasn't really concerned with subjectivity or objectivity here at all in their *Objectivity*, 17. On the emergence of the related value of "impartiality" in the seventeenth and eighteenth centuries, see Murphy and Traninger, eds., *The Emergence of Impartiality*.

[25] Bacon, *Novum Organum*, 11.110.

[26] Moshenska, *Iconoclasm as Child's Play*, x.

this process. There is also a critical difference, however. In the practice described by Moshenska, child's play is itself iconoclastic, diminishing and desacralizing once-holy objects by infusing them either with its triviality and tawdriness or with its wild anarchic energy.[27] For Bacon, however, a return to the state of childhood is celebrated as the outcome of such iconoclasm. Drawing, like Traherne, on the Gospel of Matthew, Bacon suggests that the ultimate result of the experimenter's destruction of his own inherited mental prejudices is a return to the spiritual and epistemological purity of youth—a regression on which humankind's entry into the kingdom of natural knowledge depends.

The importance of the conceptual category "innocence" to the emergence of objectivity has been brilliantly expounded by Joanna Picciotto, who however focuses not on children, but on prelapsarian Adam as its primary embodiment in the minds of seventeenth-century experimentalists. As Picciotto writes:

> The concept of objectivity suggests that we place our trust in the perspective of the innocent eye ... The question raised by objectivity is how innocence, traditionally understood to be a state of ignorance, ever came to be associated with epistemological privilege.[28]

Picciotto goes on to suggest that the association results, in her words, "from the seventeenth century's conversion of ... Adam, into a specifically intellectual exemplar".[29] Importantly, Adam was a worker, as well as the original innocent: experimentalists redeem curiosity from its connections with original sin by aligning it with "investigative labour rather than appetite", drawing on the knowledge and skills of the "unlettered" agricultural workers, tradespeople, and artisans they supposed to be Adam's successors, "free of philosophical prejudices" and committed to learning though toil.[30] In presenting Adam in this way, experimentalists reversed the traditional hierarchy of *otium*, leisure, and *negotium*, business or labour, insisting that manual industry, rather than aristocratic leisure, is the condition for learning.

---

[27] "The temporal unpredictability and disorganization of ... children" is especially important: "offering the object up to the child's temporal world is a way of severing it from the carefully ordered forms of sacred time in which it had previously been embedded". Moshenska, *Iconoclasm as Child's Play*, 191.

[28] Picciotto, *Labours of Innocence*, 1.

[29] Picciotto, *Labours of Innocence*, 1.

[30] Picciotto, *Labours of Innocence*, 3, 172.

It is undeniable that seventeenth-century promoters and practitioners of experimentalism held up the figure of prelapsarian Adam as a paradigm of sensory acuity. As Joseph Glanvill—clergyman and apologist for experimentalism—famously put it in *The Vanity of Dogmatizing* (1661), "Adam needed no Spectacles", for "the acuteness of his natural Opticks ... shewed him much of the Coelestial magnificence and bravery without a Galileo's tube".[31] It is also true that they celebrated the contributions of "*Mechanicks*", "*Merchants*", and "*Husbandmen*" to the new science, in terms which accentuated their putative sensory integrity. Such men, Sprat writes, make a crucial contribution to the Royal Society, being "plain, diligent, and laborious observers ... who, though they bring not much knowledg, yet bring their hands, and their eyes uncorrupted", and whose "Brains" are not "infected by false Images".[32]

As I have suggested, however, we may add to Picciotto's foregrounding of prelapsarian Adam and his offspring amongst the horny-handed sons of toil an alternative genealogy: the idea of a specifically childlike innocence and sensory lucidity was also key to the development of ideas about objectivity. "Those Pure and Virgin apprehensions I had from the Womb", proclaims Traherne, "and that Divine Light wherwith I was born, are the Best unto this Day, wherein I can see the Universe ... Certainly adam in Paradice had not more sweet and Curious apprehensions of the World, then I when I was a child".[33] Traherne implies that the clarity of the child's perceptions of the world at least equal—and possibly exceed—those of Adam in paradise. Indeed, there is some overlap here: in early modern England, prelapsarian Adam—following the authority of the early Christian theologian Ireneus—was sometimes thought of as childlike, and children, conversely, approximated the condition of innocent Adam.[34] As the cleric and author John Earle put it in his *Micro-cosmographie* (1628): "A Childe

---

[31] Glanvill, *The Vanity of Dogmatizing*, C3r (5).

[32] Sprat, *History of the Royal Society*, I4v (72).

[33] Traherne, *Centuries* 3.1, in *The Works*, 5.93. Jane Partner comments that, for Traherne, "the untainted vision of the child ... resembles the faculty of vision that Adam possessed before the Fall" in *Poetry and Vision*, 93. Similarly, Timothy Harrison argues that Traherne's "neonatal memories allow him to see the world as if it were Eden". Harrison, *Coming To*, 149.

[34] See Steenberg, "Children in Paradise", 1–22; and Harrison, *Coming To*, 151. In contrast, Elizabeth S. Dodd argues that "despite comparisons of the infant self to Adam ... Traherne's Adam is not the child of Irenaen theology. He is a man ..." Dodd, "'Perfect Innocency'", 220.

is ... the best Copie of Adam before hee tasted of Eve, or the Apple. ... He is purely happy, because he knowes no evill ... He is the Christians example".[35] It is no wonder, then, that in their efforts to regain the sensory perfection and mastery of nature possessed by Adam before the Fall, natural philosophers as well as poets and theologians turned to his closest "Copie[s]" on earth—children—to pursue their aims.

In this context, innocent play was just as vital as what Picciotto identifies as an emphasis on the blamelessness of prelapsarian labour.[36] Importantly for my purposes here, one way of defining play is by its freedom from extrinsic motivation, and therefore its disinterestedness. As Johan Huizinga writes in *Homo Ludens* (1938), play is "a voluntary activity ... never imposed by physical necessity or moral duty"; it is, therefore, a form of "freedom" which "stands outside the immediate satisfaction of wants and appetites".[37] Play, he confirms later in the same work, "is an activity connected with no material interest, and no profit can be gained by it".[38] Huizinga's observation has a long prehistory. As Stephen E. Kidd explains, in three of his late works—the *Sophist*, *Statesman*, and *Laws*— Plato "establishes 'play' (*paidia*) as the ... overarching category set above all forms of what we today name 'art': music, poetry, theatre, sculpture, painting, and so forth" on the basis that play is "for the sake of pleasure alone".[39] Play's supremacy derives from its independence from extrinsic motivations: as the Athenian says in the *Laws*, is a "harmless pleasure", where "there is neither loss nor profit of any serious worth".[40] In contrast, as Kidd explains, "Aristotle demotes Plato's play by making it exclusively an activity of the body".[41] For Aristotle as for Plato, play is autotelic: "the pleasurable activities of play", he claims in the *Nicomachean Ethics*

---

[35] Earle, *Micro-cosmographie*, B1r-2v (unpaginated).

[36] Whilst he does not discuss play or playfulness specifically, David Carroll Simon has recently challenged Picciotto's emphasis on labour, arguing the early modern experimentalists did not so much reverse the traditional hierarchy of *otium* and *negotium*, as insist on their interpenetration. Simon, *Light without Heat*, 18–21.

[37] Huizinga, *Homo Ludens*, 7–9.

[38] Huizinga, *Homo Ludens*, 13. See also Roger Caillois: "play must be defined as a free and voluntary activity" (*Man, Play and Games*, 6); and Hans-Georg Gadamer: play "is not tied to any goal that would bring it to an end; rather, it renews itself in constant repetition" (*Truth and Method*, 104).

[39] Kidd, *Play and Aesthetics*, 7.

[40] Plato, *Laws*, 2.667e; quoted in Kidd, *Play and Aesthetics*, 58.

[41] Kidd, *Play and Aesthetics*, 123.

(c.335–322 BC), "are not chosen for the sake of other things".[42] Although he accepts that moderate play might—like sleep—be a necessary form of recreation or rest, however, because it is concerned with "pleasures of the body" it cannot be virtuous or a central characteristic of *eudaimonia* (happiness, or the best life).[43] Rather, according to Aristotle, *eudaimonia* consists in *theōria*, contemplation.[44]

In the Middle Ages, ancient ideas about the autotelic nature of play were endorsed and disseminated by Thomas Aquinas in his *Summa Theologica* (1265–1274). In the section addressing temperance, which draws heavily on the *Nicomachean Ethics*, Aquinas affirms that "playful actions themselves considered … are not directed to an [extrinsic] end" and concludes that because "the pleasure derived from such actions is directed to the recreation and rest of the soul" (i.e. an intrinsic end) consequently "it is lawful to make use of fun".[45] In the seventeenth century, discussions of this passage often occurred in the context of debates surrounding the legitimacy of theatrical "plays", with defenders of the dramatic arts leveraging Aquinas' authority as evidence for the validity of such recreations. Richard Brathwaite, for instance, concludes in *The English Gentleman* (1631) that "repairing to *Stage-playes*" is "not altogether to be disallowed" on the basis that "that *Thomas Aquinas* giveth instance in *Stage-playes*, as fittest for refreshing and *recreating* the minde".[46] As well as enabling the guardians of drama to avow its re-creative or regenerative benefits, however, Aquinas also opened up the possibility of locating the righteousness of play understood more broadly in its freedom from external agendas and interests.[47] In particular, the perceived absence of the

---

[42] Aristotle, *Nicomachean Ethics* 10.6, 1176b9–10; quoted in Kidd, *Play and Aesthetics*, 129.

[43] Aristotle, *Nicomachean Ethics* 10.7, 1177b22; quoted in Kidd, *Play and Aesthetics*, 129.

[44] Aristotle, *Nicomachean Ethics* 10.8, 1178b8–10; quoted in Kidd, *Play and Aesthetics*, 135.0.

[45] Aquinas, *Summa Theologiæ*, Part II.II, Q.168, Art. 2. For an overview of Aquinas' ideas about play, and for the argument that Aquinas considered theology itself a form of play, see Whiddens III, "The Theology of Play". Umberto Eco also comments on this passage in *The Aesthetics of Thomas Aquinas*, 17: "Pure, disinterested contemplation is similar to play, because it is an end in itself. It also resembles play in that it is not a response to some compulsion rooted in the exigencies of life, but is rather a higher activity appropriate to a spiritual creature". On Aquinas' role in disseminating Aristotelian ideas about play in the European Middle Ages, see Olsen, "Play as Play", 197.

[46] Brathwaite, *The English Gentleman*, Aa4r-v.

[47] Aquinas, *Summa Theologiæ*, Part II.II, Q.168, Art. 2.

instrumentalism which distorts other realms of human activity means that play may have epistemological, as well as spiritual or psychological, benefits.

It is this way of thinking about play that informs Erasmus' avowal, in his immensely popular *De Civilitate Morum Puerilium*, translated as *The Civilite of Childehode* by Thomas Paynell in 1560, that:

> In honest playes there must be a certaine lustines and mirthe, so there be no sticking in opinions the whiche is the mother of debates: & that there be no disceit nor lying: for of these smal beginninges, come greater injuries and malice. He gayneth more honestly from debate, than he that obtayneth or winneth the thinge troubling himselfe by debate. ... Men muste play to refresh their spirites, and not for gain.[48]

As for Aristotle and Aquinas, the point of play is refreshment or recreation, not "gain". Yet it is just this quality of freedom, of liberation from the exigencies of external ends, that makes play productive and valuable: framing play as a kind of autotelic, low-stakes intellectual frolic, Erasmus suggests that where there is "no sticking in opinions" and no "disceit", the players "gayneth more" than they would had they seriously and single-mindedly pursed victory by any means.

Historians have shown how classical, medieval, and early modern ideas about the disinterested nature of play fed, via their reformulation in work of Immanuel Kant and Friedrich Schiller, into to modern aesthetic theory.[49] Less frequently noted is the way those ideas also informed the emerging ideology of scientific objectivity. For early experimentalists, autotelic play, as it was conceptualized in the theological and humanist pedagogical traditions, provided a way to envisage the mode of enquiry necessary for a form of natural philosophy which could become a *scientia operativa*, productive of works, only—paradoxically—through the use of (in Sprat's words) "impartial trials", driven by the intrinsic desire for knowledge rather than what Bacon calls the "use and fruit" of external worldly rewards.[50]

---

[48] Erasmus, *The Civilite of Childehode*, E4r.

[49] See Hein, "Play as an Aesthetic Concept", and Moshenska, *Iconoclasm as Child's Play*, 8–12. On Kant's interest in *naturspiel* or the play of nature specifically, see Findlen, "Ludic Postscript", 63–4.

[50] Sprat, *History of the Royal Society*, Dd4r (215); Bacon, *Novum Organum*, 11.112. On *scientia operativa*, see Klein, "Francis Bacon's *Scientia Operativa*".

Recognizing the relevance of childlike innocence and play to early ideas about the importance of impartiality in pursing natural knowledge has consequences for our understanding of the historical development of objectivity. In their ground-breaking book on the topic, Lorraine Daston and Peter Galison argue that the "epistemic virtue" of objectivity in the modern sense—the effacement of subjectivity in pursuit of accurate knowledge—did not exist prior to the nineteenth century.[51] Instead, early modern and Enlightenment virtuosos valued something different: "truth-to-nature", or knowledge of nature in its most generic, essential, universal, "perfect" form, expunged of the idiosyncrasies, variations, and deviations which so often characterize particular instances of a phenomenon. Knowledge of this kind depended not on the elimination of the knower's subjectivity (as in the case of objectivity) but rather on a kind of self-assertion: the scientist must draw on their experience and judgement to choose particular examples and combine them into an ideal type, free from the irregularities of imperfect individual specimens. "For Enlightenment savants", Daston and Galison claim, "achieving truth-to-nature required that they actively select, sift, and synthesize the sensations that flooded the too-receptive mind. ... To register experience indiscriminately was to be at best confused and at worst indoctrinated". With the rise of objectivity in the nineteenth century, however, "the subjective self of ... scientists was viewed as overactive and prone to impose its preconceptions and pet hypotheses on data". As such, "these scientists strove for a self-denying passivity".[52] In sum, early modern "truth to nature" required an active, selective, interventionalist mode of knowing; nineteenth-century "objectivity", in contrast, demanded self-abnegation and passivity.

More recently, scholars have challenged Daston's and Galison's insistence on the radical novelty of objectivity in the nineteenth century, whilst reinforcing aspects of the model of epochal change they so compellingly describe. Most relevant here is Alexander Wragge-Morley's description of what Daston and Galison call "truth-to-nature"—the exercise of judgement in the discovery of perfect types—as a "premodern image of

---

[51] "Objectivity the thing was as new as objectivity the word in the mid-nineteenth century". Daston and Galison, *Objectivity*, 34.

[52] Daston and Galison, *Objectivity*, 203.

objectivity", rather than a stark alternative to it. However, Wragge-Morley concurs with (and elaborates on) their characterization of seventeenth- and eighteenth-century science as dependent on the assertion of subjective judgement, arguing cogently that aesthetic pleasure played a decisive role in "judging the perfection of individual specimens".[53] For seventeenth-century naturalists, Wragge-Morley shows, "the belief that the world, including human responses to sensory experience, was the product of divine design", legitimated the role of "judgments of taste" in understanding the natural world, as such judgements "reflected an order emanating from the wisdom and goodness of God".[54]

Attending to moments in the seventeenth century where children are commended as exemplary observers of nature, however, offers a slightly different perspective. For a start, it cements the conclusions of scholars such as Perez Zagorin, who have argued that the modern concept of objectivity—objectivity as the elimination of subjective judgement—has deeper historical roots than Daston and Galison recognize.[55] According to Daston and Galison, "by the mid-nineteenth century, scientists ... aspired to waxlike receptivity", and this constituted one aspect of the new value placed on objectivity.[56] Similarly, women were employed "to do astronomical calculation and classification" on the basis that "the very possibility of employing 'unskilled' workers served as a tacit guarantee that data thus gathered were not the figment of a scientist's imagination or preexisting philosophical commitment ... beyond their supposed 'lack of skill,' women workers were presumed to offer a 'natural' predilection away from the grand speculative tradition".[57]

In the seventeenth century, however, children were already valued by experimentalists for precisely this quality of receptivity, and a concomitant

---

[53] Wragge-Morley, *Aesthetic Science*, 102. On the relevance of aesthetic taste to the early history of science, see also Steven Shapin, "The Sciences of Subjectivity", *Social Studies of Science* 42.2 (2011): 170–84.

[54] Wragge-Morley, *Aesthetic Science*, 162, 179.

[55] Zagorin contends that "in the Western intellectual tradition, some of the ingredients presently constituting the concept of objectivity long antedated the time of Francis Bacon and can be traced back to classical antiquity" in "Francis Bacon's Concept of Objectivity", 380.

[56] Daston and Galison, *Objectivity*, 95.

[57] Daston and Galison, *Objectivity*, 341.

lack of subjective judgement. As Hooke wrote in the manuscript now known as the "Philosophicall scribbles", dating from the early 1680s, the human sensorium is "like a peice of soft wax" which "'receive[s] ... impressions and stamps' from external objects". So far, so Aristotelian. This process is, however, expedited in the case of the child, who will "(like a young and unskillful receiver of cash) take & lay up for true and good all that comes whether they be soe or noe".[58] And although Hooke's tone is censorious here, elsewhere children are valued highly by some seventeenth-century natural philosophers (including Hooke) precisely for this lack of judgement, and for their freedom from what Daston and Galison call "preconceptions and pet hypotheses": for their inability to "actively select, sift, and synthesize the sensations that flooded the too-receptive mind". Hence, in his *Parasceve ad Historiam Naturalem* [*Preparative to a Natural History*] of 1620, Bacon instructs that a reformed "History of the Arts":

> should comprise matters so commonplace that people would imagine that, as everyone knows about them, it would be pointless to write them down. In the second place, it should comprise things vile, illiberal, and repellent (for to the pure all things are pure ...). In the third place, it should also adopt things frivolous and childish (and no wonder as we must become again quite childlike).[59]

The ecumenism of Baconian natural history, with its omnivorous interest in things "ordinary", "mean" and "trifling" is—once again—associated with a return to the state of childhood. In this respect, the mode of observing embodied by the child is closer to (if not identical with) the nineteenth-century ideal of objectivity than to the earlier virtue of "truth-to-nature": children are excellent empiricists because, lacking higher-level speculative reason or extensive experience, they haven't yet developed subjective biases to impose on nature.

The pursuit of truth about nature, then, depends not only on acts of skilled, experienced discrimination and judgement but also on a kind of suspension of such judgement, an inability or refusal to distinguish between the beautiful and the ugly, the rational and the absurd. As

---

[58] Transcription in Oldroyd, "Some 'Philosophicall Scribbles'", 17–18.

[59] Bacon, *Preparative to a Natural History*, in *OFB*, 11.465. See also *The Advancement*, *OFB*, 4.64, where Bacon claims that "meane and small things discover great, better then great can discover the small". On this passage, see Aulakh, "'Small Things Discover Great'", 53–4.

Wragge-Morley shows, for seventeenth-century naturalists understanding nature may well ultimately involve the exercise of aesthetic taste, grounded in "bodily and mental cultivation". If, however, "the inability to take pleasure … from the encounter with specimens of divine design" was "a product of human corruption", in order to reach that point the natural philosopher must first discard accrued expectations and standards.[60] They must, that is, also cultivate a kind of naïve, undiscriminating interest in the natural world: a "Virgin mind", as Charleton has it, oblivious to the fine distinctions which inform aesthetic choices.

We can see this with particular clarity if we turn to seventeenth-century cultures of collecting, where a childlike lack of discrimination and a playful pleasure in trivialities is both derided and defended. The propensity of children to esteem things which adults dismiss as insignificant is often noted today as in the early modern period; in the twenty-first century, Melissa Kaseman's photographic art series *Preschool Pocket Treasures* provides an elegant illustration of the ephemera a contemporary child counts as riches, documenting the sticks and sequins, bird's feathers and lone Lego pieces, browning cherry blossoms and burst balloons found in her three-year-old son's pockets.[61] Seventeenth-century authors, too, commented on the intense affection and sustained perceptual attention that children often give to things which adults ignore: as Ben Jonson wrote in his commonplace-book *Timber, or Discoveries* (1640), "children … esteem every trifle".[62] In moral-philosophical and devotional works, this tendency is often taken as evidence of children's underdeveloped mental capacities: "Wee will laugh at little children, to see them part with rich Jewels for childish trifles", writes the cleric John Denison in a 1621 work of spiritual instruction; "yet alasse, daily experience doth proclaime it that many are so childish, to part with these rich and precious Jewels their *Soules, for* base trifles, and … *for* earthly vanities".[63]

---

[60] Wragge-Morley, *Aesthetic Science*, 162.

[61] Kaseman, *Preschool Pocket Treasures*, http://www.melissakaseman.com/preschool-pocket-treasures. On how "young collectors" in the modern world "imitate the activities of sophisticated art collectors by acquiring, exchanging, safekeeping, and showing their items", in ways which "echo museum functions", see Stone, "Children's Collections", 77. On modern children as collectors of natural phenomena specifically, see Lekies and Beery, "Everyone Needs a Rock".

[62] Jonson, *Discoveries*, 548.

[63] Denison, *The Christians Care*, C6r (35). See also John Sheffield, *The Rising Sun* (1654), F1v (66): "How great is the folly of the sons of men, who toil, sweat, fret … to get these

Adapting the trope, satirists of early Royal Society often derided the activities of individuals such as Hans Sloane—a prolific collector of natural and antiquarian curiosities—in similar terms. In the anonymously published *An Essay in Defence of the Female Sex* (1696), probably by Judith Drake, the author characterizes "Virtuosos" primarily as obsessive collectors of useless trifles, who:

> amuse themselves continually with the Contemplation of those things, which the rest of the World slight as useless, and below their regard. ...The *Virtuoso* ... is one that has sold an Estate in Land to purchase one in *Scallop, Conch, Muscle, Cockle Shells, Periwinkles, Sea Shrubs, Weeds, Mosses, Sponges, Coralls, Corallines, Sea Fans, Pebbles, Marchasites* and *Flint stones*; and has abandon'd the Acquaintance and Society of Men for that of *Insects, Worms, Grubbs, Maggots, Flies, Moths, Locusts, Beetles, Spiders, Grashoppers, Snails, Lizards* and *Tortoises* ... He is ravish'd at finding an uncommon shell, or an odd shap'd Stone ... A piece of Ore with a Shell in it is a greater Present than if it were fine Gold.[64]

These, she continues, are "his darling Toys": the virtuoso's prized collection of trivial bits and bobs aligns him with child for whom seashells, insects, and pebbles are precious treasures, comprising a "Philosophical *Toy Shop*" or "*Raree Show*" (a portable puppet show) which he shows to observers "in a whining *Tone*".[65]

Elsewhere, however, the child's lack of discrimination is depicted in commendatory terms. In his *Centuries of Meditations*, Traherne recollects that in his youth:

---

earthly things. ... We count it childishnesse to see our boyes to run after painted Butter-flyes, wrangle for a Top, and fight for a Ball: We are the more children who pant (as if out of breath) for the dust of the Earth". The ur-text here is, again, Aristotle's *Nicomachean Ethics*: "Just as things are valued differently for children and adults, so also for bad adults and good ones". Aristotle, *Nicomachean Ethics* 10.11, 76b23–4; quoted in Kidd, *Play and Aesthetics*, 37.

[64] Drake, *An Essay in Defence*, G8v-H1r (96–7). For discussion of the authorship of *An Essay in Defence*, see Devereaux, "'Affecting the Shade'".

[65] Drake, *An Essay in Defence*, H3r (101). Perhaps ironically, Drake exempts Boyle and the Royal Society as an institution from her critique: "I us'd to say, I thought Mr. *Boyle* more honourable for his learned Labours, than for his Noble Birth; and that the *Royal Society*, by their great and celebrated Performances ... highly merited the *Esteem, Respect* and *Honour* paid 'em by the Lovers of Learning all *Europe* over. But tho' I have a very great Veneration for the *Society* in general, I can't but put a vast difference between the particular Members that compose it". Drake, *An Essay in Defence*, H4v-5r (104–5).

It was a Difficult matter to persuade me that the Tinsild Ware upon a hobby hors was a fine thing. They did impose upon me, and obtrude their Gifts that made me believ a ribban or a feather Curious. I could not see where the Curiousness or fineness: and to Teach me that a Purs of Gold was of any valu seemed impossible, the art by which it becomes so, and the reasons for which it is accounted so were so Deep and hidden to my Inexperience. ... So that nature is still nearest to natural Things.[66]

Arguing from his own experience, Traherne avers both that that childish perception is undiscriminating in the sense of non-discriminatory, and that—in their freedom from prejudicial custom and preconception—children gravitate not to artificial toys but rather to natural joys. While the lust of adults for the gaudy, artificial glitter of worldly lucre imperils their immortal souls, the tendency of children to value apparently trivial natural things which are, nonetheless, products of divine workmanship is evidence of an innate piety.

As we saw at the beginning of this book, moreover, satirists of the new philosophy often regurgitate or respond to the rhetoric employed by its advocates, and Judith Drake is no exception: figures including Robert Hooke present the child's lack of discrimination as an epistemic virtue. Thus, in his posthumously published "General Scheme, or Idea of the Present State of Natural Philosophy", Hooke expands on Bacon's comments in the *Parasceve* about the importance of "frivolous and childish" things to natural history, declaring that:

Observations, Experiments, or Circumstances are not to be esteem'd according to the common Opinion of the World, nor are they to be look'd upon as they are ... gainful, or sumptuous, or esteem'd by the great, the grave, the otherwise Learned ... He that is a true Philosophical Historian ... may perhaps see Cause to account those [things] the most Precious and Rich, which are generally esteem'd the most vile and sordid ... Those things which others count Childish and Foolish, he may find Reasons to think them worthy his most attentive, grave and serious Thoughts; and those things which some are pleased to call Swingswangs to please Children, have been found to discover irregularities even in the Motion of the Sun it self ... He ought therefore to fortify himself against these kinds of Prejudices, which are too apt to intrude into his Mind, and prepossess him against a clear View and Observation of Circumstances as they are in their own Nature.[67]

[66] Traherne, *Centuries* 3.9, in *The Works*, 5.97–8.
[67] Hooke, "A General Scheme", B1r-T1v (1–70).

For Hooke, the ignorance of children about what the "the great, the grave, the otherwise Learned" adult world perceives as valuable, and their inclination to dwell on the trivial and oft-overlooked natural things, is laudable: it is the result of "a clear View and Observation of Circumstances as they are in their own Nature", and a disposition to which "a True Philosophical Historian" should aspire.

In fact, there is some evidence that men such as Sloane and the natural historian James Petiver may have actually received contributions from children in building their spectacular collections, with Petiver commenting to a correspondent in 1695 that his instructions for collecting specimens were so clear that "any child of 6 years old" should be able to follow them.[68] Antiquarians, too, accepted or solicited offerings from children: in his *Parochial Antiquities* (1695), the antiquarian (and bishop) White Kennett describes having bought more than a hundred Roman coins, "most of which have been found by the Children of *Wendlebury* in following the plough, or by turning the clods of earth".[69] Treated for centuries as idle curiosities or playthings, for men such as Kennett such items acquired new significance as tokens with the potential to unlock the secrets of the past.

* * *

In this chapter, I have maintained that the figure of the innocent and playful child played a central role in conceptualizing and valorizing the principles of seventeenth-century experimental philosophy, most notably its promulgation of a naïve, unprejudiced, undiscriminating form of apprehension akin to something like objectivity in the modern sense. Less inclined to see the world through the obscuring gauze of cupidity,

[68] James Petiver to George Wheeler, 18 May 1695 and 29 October 1696, Sloane MS 3332, fols. 124, 223–5; quoted in Delbourgo, *Collecting the World*, 206. On the contributions made both by "unnamed slave women", who "gathered specimens for colonial collectors who sent them back to patrons like James Petiver and Sloane", and by "the wives of merchants and planters", see Delbourgo, *Collecting the World*, 203.

[69] Kennett, *Parochial Antiquities*, B3r (13). In the Preface to his account of "the natural rarities of England, Scotland, & Wales", the Baconian virtuoso Joshua Childrey invites "young Gentlemen to look and search into these pleasant Speculations more then heretofore they have done, and to visit each his neighbour cuosities [*sic*]." Childrey, *Britannia Baconica*, B3v, B4v. On children's contributions to antiquarianism in the long nineteenth century—including their role in preserving the "trifles" that antiquarians reconfigured as "relics"—see Rowland, *Romanticism and Childhood*, Chapters 4, 5, and 6.

children were more inclined to attend intensely yet indiscriminately to natural phenomena. The rhetoric of childness, besides, also highlights an inconsistency in objectivity as it has been conceived and practiced in later eras. The nineteenth-century scientists described in Daston's and Galison's book must *strive* for "a self-denying passivity"; perversely, the self-abnegation objectivity demands are founded in an act of will, on the individual's volitional and active suppression of individuality. Similarly, according to Picciotto, innocence-cum-objectivity can only be reconstituted as a condition of natural philosophical knowledge via its association with strenuous (Adamic) labour. The impartiality of the child, however, is unconscious, effortless, and spontaneous: there is no need to supress preconceptions which have not yet been acquired. Playful and pleasurable, rather than arduous and self-denying, the distinctive form of impartiality which early modern thinkers attribute to tender youth is not merely useful and virtuous; it is also fun. In the next chapter, I explore the playfulness of experiment in more detail, focusing in particular on the new technologies of the microscope and telescope.

CHAPTER 4

The "Toyish Art" of the Microscope: Experiment as Play

Abstract This chapter investigates the ways in which natural philosophers including Robert Hooke attempted to emulate or recreate a childlike perspective in their own experimental activities. For members of the early Royal Society, both the playfulness and the sensory lucidity of childhood could be partially recaptured in adulthood through the use of prosthetic aids. In particular, the telescope and microscope were—for all their technological complexity—sometimes understood in a very literal way as toys which nonetheless served a vital function, promising to restore something of this lost childlike sensibility. This effort, I suggest, was underpinned by a transformation in ideas about experience as a source of knowledge, as an Aristotelian model of "experience" as conferring understanding only gradually, over an extended period of time (and hence, as the exclusive preserve of the old), was replaced by a new, Baconian emphasis on the heuristic and probative value of experimental intervention in the present moment (a form of experience equally available to children).

Keywords Children • Science • Natural philosophy • Experiment • Royal Society • Play • Robert Hooke • *Micrographia* • Robert Boyle • Francis Bacon • Microscopes • Telescopes • Experience • Vision • Toys • Margaret Cavendish • Samuel Butler

© The Author(s), under exclusive license to Springer Nature Switzerland AG 2024
E. L. Swann, *Science as Child's Play in Seventeenth-Century England*, https://doi.org/10.1007/978-3-031-75849-2_4

In a poem included under the entry for "Ant" in his *Commentaries of Heaven*—an ambitious, and ultimately unfinished, encyclopaedia of reflections on a range of phenomena—Thomas Traherne enthuses that "Bright Apprehensions and Angelical / Make a Sublime Thing of a very small".[1] What Jane Partner calls Traherne's "fascination with the wondrous material intricacies revealed by microscopy" is well-established, and whilst the wider context of Traherne's work, with its glorifications of youthful vision, encourages us to connect the "Bright Apprehensions" in question to a child's-eye view, in this context they also seem to refer to the observations recorded in a work Traherne very much admired: Hooke's *Micrographia*.[2] As a compendium of microscopic observations of artificial and natural objects including seaweed, seeds, bird feathers, moss, and moths wings—and, indeed, ants—*Micrographia* shows Hooke practising what, in the later "General Scheme", he preached, taking as the objects of his serious attention "things which others count Childish and Foolish".[3] It is a key premise of this chapter that in *Micrographia*, Hooke is—by his own lights—childlike in his preoccupation with what, in "To the Royal Society", Abraham Cowley calls "the things which ... proud men despise, and call / Impertinent, and vain, and small".[4]

Despite this, at first glance the importance of sophisticated optical technologies—notably the microscope and telescope—to seventeenth-century natural philosophy may seem to pose a challenge to my argument so far. The range and inconsequentiality of the objects Hooke scrutinizes in *Micrographia* might recall the indiscriminate interests of the juvenile naturalist discussed in the previous chapter, but the engravings of them that he produces are not quite, as he claims, "the things themselves as they appear" to the naïve observer, in all their quirky singularity; rather, they are aggregates of repeated acts of scrutiny and products of "*mature deliberation*".[5] "I never began to make any draughts", Hooke explains, "before many

[1] Traherne, *Commentaries of Heaven*, in *The Works*, 4.95.
[2] Partner, *Vision and Poetry*, 96. This fascination, Partner shows, was "ambivalent ... because whilst Hooke's observation of insects revealed the exquisite details of their material forms, it did nothing to enquire into their spiritual significance". On Traherne and Hooke, see also Murphy, "No Things but in Thoughts", 60.
[3] Hooke, "A General Scheme", B1r-T1v (1–70).
[4] Cowley, "To the Royal Society", B3r (unpaginated).
[5] Hooke, *Micrographia*, a1v, b1v.

examinations in several lights, and in several positions to the light, I had discover'd the true form".[6] Hooke's images, that is, are the result of multiple observations, purged of anything extraneous and fused together to create a single representative image.[7] In this regard, they might be understood as characteristic of a period when, as Daston and Galison argue, "truth to nature", rather than "objectivity", was the dominant value. If natural philosophers valued the untutored senses and playful instincts of the child so highly, moreover, why did they invest so much in the ability of these complex instruments, the creation and operation of which requires laborious training and skill, to mitigate their own failings as visual observers?[8] The driving conviction of *Micrographia*, after all, is that the unassisted, guileless eye can confer only superficial knowledge of the natural world: "we must", pronounces Hooke, add *"artificial Organs* to the *natural*" if we are to fully appreciate the vibrant, intricate, multitudinous miracle of creation.[9]

Statements such as this underwrite a revisionary strand, in recent historiography, which represents early experimental science not as a shift away from bookish authorities and abstract reasoning towards empirical experience, but rather as a rejection of one form of experience in favour of another. As Peter Dear, Mary Thomas Crane, and Barry Allen, amongst others, have suggested, the traditional Aristotelian natural philosophy which had held sway for so long placed a high value on quotidian, unmediated, easily verifiable, and oft-repeated observations, often encoded in axioms ("the sun rises in the east", for example).[10] As such, as Crane asserts, it had its "epistemological basis ... in ordinary perceptual experience of the world".[11] Importantly for our purposes here, experience in this sense was also gradual and cumulative, amassed over the course of a

---

[6] Hooke, *Micrographia*, f2v.

[7] Ian Lawson comments that "with a few exceptions, Hooke's published images collapsed his 'many examinations' into one naturalistic and lifelike image of a subject, and he sometimes deliberately occluded the fact that they were mosaics constructed from many views". Lawson, "Crafting the Microworld", 35.

[8] In the words of Ofer Gal and Raz Chen-Morris, "instruments embedded sophisticated mathematical knowledge and fine artisanal skill". Gal and Chen-Morris, *Baroque Science*, 79.

[9] Hooke, *Micrographia*, a2r.

[10] This is Peter Dear's example in "The Meanings of Experience", 109.

[11] Crane, *Losing Touch with Nature*, 4.

lifetime. As Aristotle writes in his *Posterior Analytics*, it should be understood not as immersion in the present moment, but rather the recollection and synthesis of multiple perceptions: "From perception there comes memory ... and from memory (when it occurs often in connection with the same thing), experience; for memories that are many in number form a single experience".[12] In Allen's paraphrase, experience for Aristotle "is not just a moment of conscious perception; it is having learned from perception. Experience is not something *had*, it is something recalled, a deferred, belated perception, a quality of the remembered past".[13] For Aristotle and his followers, then, "experience" does have a specific, heuristic function within natural philosophy—but this kind of experience is not fully distinguished from "experience" understood more broadly and loosely, as the knowledge and wisdom that comes with age.

Bacon and his followers, on the other hand, insisted that experience must take the form of its close cognate, experiment, in order to be optimally useful to the natural philosopher, who must now intervene in the natural world in order to understand it.[14] As Barry Allen writes, "what led to the break with scholastic natural philosophy was not authority finally bending to experience; it was a new kind of experience, experience under controlled, highly artificial conditions—in a word, *experimental* experience".[15] In contrast to the learned men of the past, who, Bacon says in the *Novum Organum*, relied on "loose and vague observation" and "unguided experience" and who were consequently merely "groping in the dark", the new natural philosopher recognizes that "nature's secrets betray themselves more through the vexations of art than they do in their usual course", and therefore that "hope of further advancement of the sciences will be well grounded only when we take and gather into natural history many experiments".[16] Henceforth, scientific experience would not

---

[12] Aristotle, *Posterior Analytics* 2.19, 100a 4–9; quoted in Dear, "The Meanings of Experience", 122. See also Aristotle, *Metaphysics* 2.1552: "from memory experience is produced in men; for many memories of the same thing produce finally the capacity for a single experience".

[13] Allen, *Empiricisms*, 7.

[14] The close etymological and conceptual connection between "experience" and "experiment" in the early modern period is well-known. As Peter Dear explains, "the Latin words generally used to denote 'experience' in both the medieval and the early modern periods, *experientia* and *experimentum*, were generally interchangeable, with no systematic distinction between them except in particular contexts". Dear, "The Meanings of Experience", 106.

[15] Allen, *Empiricisms*, 147.

[16] Bacon, *Novum Organum*, 156–60.

be passively received but actively sought and artificially orchestrated, mediated by complex methodologies and technologies, obscure to the uninitiated perceiver. As Allen puts it, "common sense and untutored perception reveal nature's outer appearance only … Nature's secrets … are hidden beyond the reach of ordinary perception, and have to be hunted out by extraordinary means".[17] In a departure from the Aristotelian understanding of experience as something slowly accumulated over time, the understanding of experience embedded in this epistemology also gives more prominence to momentary, transient, exceptional sensations and perceptions.[18]

The broad historical narrative represented here by Dear, Crane, and Allen represents a significant breakthrough in our understanding of the origins of modern science, nuancing an older, more straightforwardly celebratory story (originating in part with members of the early Royal Society themselves) about the replacement of crabbed, abstruse, dogmatic scholasticism by the clear, cogent, liberating light of empiricism. Nonetheless, it misses an essential paradox lying at the heart of seventeenth-century experimentalism, namely the extent to which mastering its sophisticated methods is both predicated on and enables a return to the guilelessness of childhood. From one perspective, to be sure, the new emphasis on the experimental manipulation of nature—and its attendant technologies— can be seen as rejection of naïve experience. As Crane argues, "the new science destroyed a direct, intuitive connection with nature".[19] Because the promotion of experiment also depended on the elevation of a form of experience grounded in the immediacy of present sensations rather than recollection of past perceptions, however, it also cleared the ground for a revaluation of the forms of knowledge available to children. What Michael Witmore describes as "the presentism of child consciousness" was

---

[17] Allen, *Empiricisms*, 152.

[18] Adam Rzepka traces a tension between experience as slowly accumulative and retrospective, and experience as the revelatory immediacy of the present moment, in Shakespearean theatre in "Rich eyes and poor hands". The two types of experience are distinguished in the German language, as Martin Jay explains. *Erlebnis*, a transitive verb, "generally connotes … [an] immediate, pre-reflective, and versional variant of experience", and "can suggest an intense and vital rupture in the fabric of quotidian routine". *Erfahrung*, by contrast, connotes "a more temporally elongated notion of experience based on a learning process", which "activates a link between memory and experience". Jay, *Songs of Experience*, 11.

[19] Crane, *Losing Touch with Nature*, 6.

well-established in the early modern period, noted by figures including Bartholomeus Angelicus, according to whom (in Stephen Batman's 1538 translation) "children ... thinke onely on things that be, and regard not of things that shall be".[20] Or, as William Kempe put it in *The Education of Children* (1588), "youth is is forgetfull, not greatly moved with regard of things past, or things to come, but wholy caried away with that which is before their face".[21] Whilst the cumulative form of "experience" which formed the bedrock of Aristotelian natural philosophy could only come with the passage of time and onset of age, the experimental form of "experience" valued by the early Royal Society—experience as a brief, intense interaction with the unknown and unfamiliar, untethered from past experience and preconceptions—was available to old and young alike.[22]

We can see this reappraisal of experience in Robert Boyle's *Occasional Reflections*. Describing the benefits of "attentive observation", Boyle writes that "attention, like a magnifying glass, shews us, even in common Objects, divers particularities, undiscerned by those who want that advantage". Consequently,

> this exercise of the mind must prove a compendious way to Experience, and make it attainable without grey-hairs; for that, we know, consists not in the multitude of years, but of observations ... nor is there any reason, why prudence should be peculiarly ascrib'd to the Aged, except a supposition that such persons, by having liv'd long in the World, have had the opportunity of many and various occurrences to ripen their judgment; so that if one man can by his attention make, as well he may in a small compass of time, as great a number of Observations as less heedful Persons are wont to do in a longer, I see not why such a man's Experience may not be equal to his, that has liv'd longer.[23]

Dilating on Bacon's dictum, in his essay "Of Youth and Age", that "a Man that is *Young in yeares*, may be Old in Houres, if he have lost no Time", Boyle decisively detaches age from "Experience". Both the quality and the quantity of "observations" matter to Boyle, but youth is no

---

[20] Angelicus, *Batman uppon Bartholome*, O1r (73); Witmore, *Pretty Creatures*, 35–6.

[21] Kempe, *Education of Children*, F1r.

[22] On the reconfiguration of experience in the seventeenth century, including by Bacon, see Jay, *Songs of Experience*, 31–7.

[23] Boyle, *Occasional Reflections*, C6v-C8r (28–31).

disqualification for obtaining either: in attending more closely to "common objects" than is usual, it is possible even in "a small compass of time" to develop the "prudence" more usually associated with the elderly, even without "grey-hairs".

From this perspective, optical technologies such as microscopes and telescopes can be understood not as a means of supplementing or improving on naïve or childish ways of experiencing natural phenomena, but rather as sophisticated tools which enabled a *return* to such ways of experiencing. Thus, when Hooke asserts in *Micrographia* that "the Science of Nature has been … too long made only a work of the *Brain* and the *Fancy*" and encourages a "return to the plainness and soundness of *Observations* on *material* and *obvious* things", warning the reader not to expect "from me any infallible Deductions" on the basis that "those … are above my weak Abilities", he aligns himself with the child whose inability to engage in higher-level reasoning becomes an epistemological advantage, preventing him from engaging in premature conjecture and freeing him to focus on the "Observations of my eyes".[24] And when he gushes that, "by the help of Microscopes … there is a new visible World discovered to the understanding", so that "the *Earth* it self, which lyes so neer us, *under our feet*, shews quite a new thing to us", he celebrates the ability of optical instruments to inspire their users with some of the shock and wonder that infuses a child's encounter with a world that is, to them, still "new"—an effect enhanced by his use of present tense, which roots the microscope's revelations in the immediate moment.[25]

Aided by the microscope, moreover, the experimenter may see with something like the acuity of children, whose vision was, if not quite prelapsarian in its perfection, certainly closer to that of unfallen Adam and Eve than that of their elders. Anticipating that "by the addition of such *artificial Instruments* and *methods*, there may be, in some manner, a reparation made for the mischiefs, and imperfection, mankind has drawn upon it self", Hooke refers both to the "errors" that arise from the "deriv'd corruption" of original sin and to those which develop as a result of "breeding and converse with men". The return to sensory perfection which the microscope "in some manner" promises goes hand-in-hand

---

[24] Hooke, *Micrographia*, b1r.
[25] Hooke, *Micrographia*, a2v.

with a stripping back of adult enculturation, the solecisms engrained by education and social life.[26]

Optimism about regaining the perspicacity of youthful vision also infused accounts of more idiosyncratic optical innovations, too. In a "letter concerning an optical experiment, conducive to a decayed sight", published in the *Philosophical Transactions* in 1668, the anonymous author—identified only a "worthy person" of "not more than 60 years of age"—describes a device which he claims has partly restored his vision, namely a pair of spectacles with the glass removed and leather tubes affixed to the frames. This device, he alleges, has had near-miraculous effects:

> I see now by this trifle (these Taper-tubes) as well as the youngest in my Family, and can read through them the smallest and blindest Prints, as ever I could from my childhood, though my fight be almost lost. And having used these empty holes for Spectacles little more than a week, I can now use them without trouble all the day long; and I verily believe, that by this little use of them, my sight already is much amended. For I now see the Greenness of the Garden, and Pastures in a florid verdure.[27]

In renewing a level of visual keenness previously lost to him since "childhood", the "Taper-tubes" enable their user both to read small print and to appreciate once again the marvels of nature. In particular, the letter-writer's rejoicing in his reinstated ability to perceive "the Greenness of the Garden" is surely not incidental: luxuriating in the verdant landscape, he celebrates the restoration of something approximating to the Edenic eye-sight to which Hooke also aspired.

The letter-writer's description of the taper-tubes as a "trifle"—a modesty topos swiftly nullified by his effusive account of their miraculous effects—is also significant, as the term's association with toys implies the childlike pleasure they confer. In the latter half of the seventeenth century, microscopes and telescopes, too, are sometimes described by both detractors and supporters of the new experimental philosophy as toys or as toy-like objects which encourage playful forms of engagement. Notably, in her *Ground of Natural Philosophy* (1668), Margaret Cavendish derides those "Human Creatures" who "trouble themselves with poring and peeping

---

[26] Hooke, *Micrographia*, a1r

[27] Anon, "An Extract of a Letter", 728–29. Pepys refers to this letter in his diary entry of 31 July 1668: "My eyes being now past all use almost. ... I am mighty hot upon trying the late printed experiment of paper Tubes". *Diary*, 9.270; see also Hooke, *Diary*, 102.

through Telescopes, Microscopes, and the like Toyish Arts"—a description that may be profitably glossed by reference to in her slightly earlier *Observations upon Experimental Philosophy* (1666).[28] Here, Cavendish objects to the work of "experimental philosophers" on the basis that:

> Artificial things are pretty toys to imploy idle time ... and though Nature takes so much delight in variety, that she is pleased with them, yet they are not to be compared to her wise and fundamental actions: for, Nature being a wise and provident Lady ... is very industrious, and hates to be idle, which makes her imploy her time as a good Huswife doth, in Brewing, Baking, Churning, Spinning, Sowing, &c ... but her artificial works are her works of delight, pleasure and pastime: Wherefore those that imploy their time in Artificial Experiments, consider only Natures sporting or playing actions.[29]

For Cavendish, it is the artificiality of microscopes and telescopes that align them with "sporting or playing": in modifying and refashioning nature, rather than simply observing how "she" behaves when left to her own devices, experimenters neglect the profitable lessons offered by her "wise and fundamental actions" in favour of an indolent fixation on childish frivolities.

Similarly framing the use of optical technologies as a form of play, Samuel Butler's satirical poem "The Elephant in the Moon", written around the early 1670s, nonetheless provides a slightly different perspective.[30] In the poem, members of a "learn'd *Society*" mistake a mouse trapped in their telescope for the lunar pachyderm of the poem's title.[31] As

---

[28] Cavendish, *Ground of Natural Philosophy*, Pp3v (294). Hilda L. Smith observes a shift in Cavendish's attitude to the microscope across the course of her work: "at first she saw it as an instrument with important potential, but in her later writings she was apt to consider it more appropriate to child's play". Smith, "Margaret Cavendish and the Microscope as Play", 40.

[29] Cavendish, *Observations*, Dd1r-V (101–2). Jonathan Shaheen comments of this passage that "Cavendish does not offer an explicit interpretation of what she means when she calls artificial things 'toys' produced by 'nature's sporting or playing actions' ... [but] it is clear from her texts that Cavendish connects nature's playing with human playing". Shaheen, "A Vitalist Shoal in the Mechanist Tide", 112.

[30] For the composition of the poem, see Quehen, "Butler, Samuel".

[31] Butler, "The Elephant in the Moon [in Short Verse]", l.1; further line references given in-text. As Robert Thyer explains in the notes to his 1759 edition of Butler's works, Butler composed two versions of the poem: one in "short Verse", another in "long". To my mind, what Thyer calls the "many considerable Additions and Variations" of the long version

they attempt to compose a "*Memorial*" celebrating their supposed discovery, their young servants take advantage of their distraction:

> The Footboys, for Diversion too,
> As having nothing else to do,
> Seeing the *Telescope* at leisure,
> Turn'd *Virtuosos* for their Pleasure;
> Began to gaze upon the *Moon* ...
> When one, whose Turn it was to peep,
> Saw something in the Engine creep;
> And, viewing well, discover'd more,
> Than all the Learn'd had done before.
> Quoth he, a little Thing is slunk
> Into the long star-gazing Trunk ...
> This being overheard by one,
> Who was not so far overgrown
> In any virtuous Speculation,
> To judge with mere Imagination ...
> He found, a *Mouse* was gotten in
> The hollow Tube ... (1.331–344)

Treating the telescope as a source of "Pleasure" and "Diversion", rather than a tool for the avid pursuit of knowledge, the footboy makes better use of it than his "Learn'd" masters: although he doesn't make any spectacular celestial discoveries, he at least uncovers the ridiculous error that, "impatient to know" (1.41), the virtuosos have manufactured. As Erik J. Jarvis points out, "speed to publication and concomitant potential for notoriety drive the Society's procedure within the poem": their error arises, in part, because of their appetite for novelty.[32] In contrast, the footboy's more dilatory, recreational use of the telescope ultimately leads to the exposure of their mistake. By the same token, it is one of the society's more youthful members who confirms the footboy's suspicions: "one/ Who was not ... overgrown" in "Speculation", and therefore able to puncture the imaginative excesses of his elders.[33]

---

detract from the terse energy and momentum of the poem; as such, I cite the shorter version. On the two versions, see Thyer, *The Genuine Remains*, 1.26.

[32] Jarvis, "The Royal Society, Collective Vision", 138.

[33] Jarvis comments that the youth of the member means he is not yet "fully inculcated into the Society's imaginative worldview", in "The Royal Society, Collective Vision", 135.

As Joanna Picciotto argues, "Butler's preoccupation with human error and its causes, [and] the intentness of his focus on the conditions under which assisted observation could go awry, make his satires virtual extensions of Royal Society propaganda".[34] And indeed, in "The Elephant in the Moon" the target of his satire is not the new forms on natural philosophy associated with the Society in toto, but the gap between principle and reality: the inability of specific members to live up to their own precepts, or specific aspects of their practices.[35] In this case, it is both the alacrity and the over-seriousness of "the Learn'd" that is lampooned: their earnest, single-minded search for new knowledge is ultimately shown to be inferior to the more (ironically) leisured, aimless, playful engagement of their servants.

Once again, satires of the Royal Society usefully illuminate an association also drawn by its fellows. In a "Discourse" read before the Society in February 1691/1692 and published posthumously in 1726, Hooke bemoaned "the Fate of Microscopes", commenting that with the exception of their devoted "Votary" Anthonie van Leeuwenhoek, "none … make any other Use of that Instrument, but for Diversion and Pastime".[36] Elsewhere, however, we also find both telescopes and microscopes extolled as toy-like by those enraptured by the pleasures and possibilities of experimental science—including, as we shall see, Hooke himself. Samuel Pepys' account, in his diary, of purchasing and using a microscope and scotoscope, has the tone of a boy both baffled and delighted by the acquisition of a much-desired plaything: "a most curious bauble it is", he commented on 13 August 1664, "and a curious curiosity it is to [see] objects in a dark room with. Mightly pleased with this".[37] For Pepys, both the microscope itself and the objects it reveals are curiosities, sources of pleasure and perplexity simultaneously.

As with the rhetorical importance of innocence, the centrality of play and playfulness to early modern natural philosophy has been noted by

---

[34] Picciotto, *Labours of Innocence*, 245.

[35] Ken Robinson, for instance, argues that Butler targets not the "new science" per se, but rather the work of "particular scientists". Robinson, "The Skepticism of Butler's Satire", 4. Jarvis, meanwhile, focuses on Butler's suspicion of the "forms of mediation", especially "publication and association", they relied on to disseminate their work. "The Royal Society, Collective Vision", 143.

[36] Hooke, "Discourse concerning Telescopes and Microscopes", S3r (261).

[37] Pepys, *Diary*, 5.240. "Bauble" was a common synonym for toy or plaything; see "Bauble, n.", OED *Online*, www.oed.com/view/Entry/16308.

some scholars. In particular, Paula Findlen has drawn attention to how in the sixteenth and seventeenth centuries natural history, and to some extent natural philosophy more broadly, "rediscovered its capacity for playfulness in the form of the scientific joke".[38] Findlen demonstrates, amongst other things, how "the play of science evolved from the play of nature ... man's ability to play mimicked nature's own instincts".[39] As with Picciotto's elision of childhood in her examination of innocence, however—discussed in the previous chapter—the kind of playfulness Findlen focuses on is adult and sophisticated, influenced by the humanist interest in *serio ludere* as a literary and philosophical practice, and she pays little attention to the play of children. In addition, Findlen implies in a number of places that the fun of natural history fizzled out with the formation of the Royal Society, arguing that "natural philosophers like Robert Boyle sought to correct the perception that nature was allowed any independent action in the process of creation ... *lusus* [games, sports, jokes] testified to the existence of an animated nature, at odds with the late seventeenth- and eighteenth-century emphasis on God as a creator of absolute categories, impermeable to the changes that a playful nature wrought".[40] Likewise, Hooke's micrographic enquiries "systematically reinterpreted an array of extraordinary phenomena, insisting that there was no play in any of these acts of nature".[41]

It is certainly true that, as Findlen argues, members of the early Royal Society were anxious to demarcate their pious project of understanding and controlling divinely created nature from the ersatz entertainments associated with the natural magic tradition, and Findlen has brilliantly illuminated their efforts in this regard. In the words of Boyle, "the works of God are not like the Tricks of Juglers or the Pageants that entertain Princes".[42] It is not quite the case, however, that they "attempted to

[38] Findlen, "Jokes of Nature", 292.

[39] Findlen, "Jokes of Nature", 315–16. This tradition, as Findlen notes, reaches back to Pliny, who in his *Natural History* described how "nature, in her ingenuity, has created all these marvels ... as so many amusements to herself", and who singled out shellfish and flowers as particular examples of nature's "sportive mood". Pliny, *Natural History*, 2:135 (VII.3), 428 (IX.52); 4:304 (XXI.1); quoted in Findlen, "Jokes of Nature", 296.

[40] Findlen, "Jokes of Nature", 325–6.

[41] Findlen, "Between Carnival and Lent", 264.

[42] Boyle, *Some Considerations Touching the Usefulnesse*, H3r (51); quoted in Findlen, "Between Carnival and Lent", 263.

purge their interpretations of nature of any traces of the ludic", as she suggests.[43] And in fact, in a later piece Findlen herself revises some of her earlier conclusions in this regard, acknowledging that—despite challenges from figures including Bernard Le Bovier de Fontenelle—"at the dawn of the nineteenth century *lusus naturae* was—and still is—a resilient category".[44]

Indeed, a sense of nature's playfulness runs like a thread through the work of later seventeenth-century experimentalists and naturalists, embellishing rather than detracting from divine craftsmanship. "We see Nature doth *ludere,* as it were, sport itself, the minute Ramifications of all the Vessels, Veins, Arteries, and Nerves infinitely varying in individuals of the same Species", writes John Ray in *The Wisdom of God*.[45] "Nature sometimes sports her self in the colours of this Bird", observes his friend Francis Willughby in his *Ornithology*, describing variations in the tail of the bird known as the whin-chat.[46] Even the famously austere Boyle was not immune to the lure of nature's jokes. In a discussion of how "volatile Salts … fasten themselves to the Receiver in various figures" during distillation, for instance, Boyle recounts the following anecdote:

> I remember that not long since subliming some volatile Salt of Urine, it adher'd to the upper part of the vessell in figures, much [like] Hartshornes … unless wee will merrily say, that the man whose urine was distill'd, had hornes given him by his wife, wee must acknowledge that nature seemes to give her selfe liberty to play in the Configuration of volatile Salts.[47]

Boyle's mischievous—if slightly convoluted—witticism implies that because the salt of distilled urine forms a branching figure that reminds him of deer horns, the person who produced the urine might be a cuckold. This is clearly quite an adult joke; nonetheless, it offers a clear illustration of the dynamic that Findlen suggests was temporarily abandoned by

---

[43] Findlen, "Between Carnival and Lent", 262.

[44] Findlen, "Ludic Postscript", 62.

[45] Ray, *The Wisdom of God*, M3r (165).

[46] Willughby, *The Ornithology*, Hh1v (234). In the same work, variations in the claws of cranes are attributed to "Nature … sporting it self, and not observing constantly the same rule", Uu1v (332).

[47] Boyle, *Some Considerations Touching the Usefulness*, Xx2r (355).

Boyle and his peers, as Boyle's own ludic instincts are stimulated by what he describes as the playful volition of nature in forming the salts.[48]

Most significantly here, in *Micrographia* Hooke explicitly depicts the challenge of viewing the celestial bodies through the telescope as a source of delightful sport. Discussing the phenomenon whereby the heavenly bodies appear either to alter in colour, "so as sometimes to appear red, sometimes more yellow, and sometimes blue", or to "twinkle" with more vigour "the neerer they appear to the Horizon", Hooke describes how:

> in examining some very notable parts of the Heaven, with a three foot Tube, me thought I now and then, in several parts of the constellation, could perceive little twinklings of Starrs, making a very short kind of apparition, and presently vanishing, but noting diligently the places where they thus seem'd to play at boepeep, I made use of a very good twelve foot Tube, and with that it was not uneasie to see those, and several other degrees of smaller Starrs, and some smaller yet, that seem'd again to appear and disappear.[49]

Here, Hooke not only attributes mischievous agency to the diminutive stars which seem to play at "bopeep" (that is peekaboo) with him, but he also accepts their invitation to the game, swapping his "three foot Tube" for a larger telescope in order to better witness their appearing and disappearing.[50]

All this is not to say that the use of optical technologies was all fun and games, however: instruction was as important as recreation. In William Simpson's *Philosophical Dialogues Concerning the Principles of Natural Bodies* (1677), Hydrophilus, a "grey headed" defender of Aristotle, complains of the discomforting experience of being schooled by youthful new philosophers: "it was never a good world since such a young fry of Novel Philosophers peep'd up", he complains, and "it's a hard case, that we must

---

[48] Boyle hints at nature's playfulness elsewhere in his works, too. In the Preface to his *New Experiments and Observations Touching Cold*, for instance, he describes how, in a "Trial, with an unseal'd Weather-glass ... a Pillar about an Inch long" of water "was kept suspended, and play'd as well conspicuously as nimbly up and down in the Pipe". Boyle, *New Experiments*, C5r (unpaginated); see also D6r (43).

[49] Hooke, *Micrographia*, Gg1v (218).

[50] It is worth noting that astronomy had long been pursued by means of play, with board games based on Ptolemaic astronomy present in Europe since the Middle Ages. See Moyer, "The Astronomers' Game". Such games, however, were highly sophisticated and complex, and probably aimed primarily at an adult audience.

be compelled to turn Schoolboys again, and go with Satchels on our backs, to learn at your Pyrotechnical and Mechanical Schools".[51] Indeed, the edification offered by the microscope was frequently represented, by Hooke and others, as a form of schooling which recalled and replicated the child's experience of mastering language.[52]

Bacon's conviction that experience must be "literate" in order to be useful is relevant here, requiring the experimentalist to become a student once again.[53] As he writes in his *Natural and Experimental History [Historia Naturalis et Experimentalis]*:

> We should beg men again and again to set aside for a while or at least discard these fickle and wrong-headed philosophies, which have put theses before hypotheses ... and to read through with due humility and reverence the volume of creatures, and dwell and reflect on it, and, purged of opinions, to study it with a pure and honest mind. This is that speech and language which went out to the ends of the earth, and did not suffer the confusion of Babel; let men learn this thoroughly and, becoming childlike, return to infancy again and deign to take its abecedaria into their hands ...[54]

We will be able to read the book of nature—a book written in the Adamic language which predates the fall of the Tower of Babel—Bacon maintains, only once we have again become "childlike" and pure of mind. Thereby restored to epistemological innocence and "purged of opinions", we might begin, laboriously and like children, to parse its Adamic alphabet.

---

[51] Simpson, *Philosophical Dialogues*, B3v (6), B1v (2), B3r (5). "Hydrophilus" is an avatar for the physician Robert Wittie; his interlocutor Pyrophilus, a staunch defender of the new philosophy, is a mouthpiece for Simpson himself.

[52] As Picciotto observes, "the fact that they were not easy to use—requiring patient application and a type of visual 'literacy' to produce reliable knowledge—helped to make optical instruments the most celebrated emblems of *reading* nature as opposed to merely gazing at it". Picciotto, *Labours of Innocence*, 214.

[53] What exactly Bacon meant by literate experience, or *experientia literata*, is contested. For a penetrating discussion, see Weeks, "The Role of Mechanics", 162–72.

[54] Bacon, *Natural and Experimental History*, in *OFB*, 12.11. See also *A New Abecedarium of Nature*, in which Bacon asserts that the work "is like some child's copybook exercise ... for it carries us forward into the kingdom of man which, in relation to the sciences, is like the Kingdom of Heaven: none enters it except in the likeness of a little child". Bacon, *A New Abecedarium of Nature* [*Abecedarium Novum Naturae*], in *OFB*, 13.172–3.

Similarly, in his *De Augmentis Scientiarum* (1623)—a Latinized and much-expanded version of *The Advancement of Learning*, translated into English in 1640—Bacon criticizes the "Logicians" for their complacent reliance on "Particulars" which are "grosse and palpable" (that is, on the obvious evidence of experience unsought) and their refusal to actively search out "Instance[s] Contradictory". This "Forme of Induction" Bacon calls "vitious and incompetent", unable to produce certain knowledge but only "probable Conjecture[s]". This inadequate and damaging way of proceeding, which has parallels in theology, is attributed to a prideful reluctance to embrace a childlike state of continual learning: "For as in the apprehending of divine truth, men cannot endure to become as a child; so in the apprehending of humane truth, for men, come to yeares, yet to read, and repeate, the first Elements *of Inductions*, as if they were still children; is reputed a poore and contemptible imployment".[55] For Bacon, the ultimate aim may be an enhancement of interpretative prowess and an increase of expertise which elevates the experimenter above the naïve, quotidian forms of experience on which Aristotelian natural philosophy relied. But getting there, paradoxically, requires a regression to the humility of infancy: a state unsullied by "opinions", and receptive to the repetitive tedium ("read, and repeate") necessary if we are to learn how to interpret the world.

In *Micrographia*—literally, "small writing"—Hooke describes his observations of a range of natural and artificial phenomena in just these terms. Descanting on the "curiously compounded shapes" of bodies viewed under the microscope, he speculates that "the Creator may, in those characters, have written and engraven many of his most mysterious designs and counsels, and given man a capacity, which, assisted with diligence and industry, may be able to read and understand them".[56] Nature's book is divinely inscribed in minute letters. Announcing his intention to begin "with the Observations of Bodies of the most *simple nature*", therefore, Hooke compares the process of learning to see through the microscope with the (re)acquisition of literacy: "we must first endevour to make *letters*, and draw *single* strokes true, before we venture to write whole *Sentences*". Subsequently, he opens the work with observations of "*Physical point*[s]" of various kinds, including "the mark of a *full stop*, or

---

[55] Bacon, *Of the Advancement and Proficience of Learning* [*De Augmentis Scientiarum*], EE4r (223).

[56] Hooke, *Micrographia*, X7v (254).

period"—an activity which soon leads him to examine "certain pieces of exceeding curious writing … one of which in the bredth of a *two-pence* compris'd *the Lords prayer, the Apostles Creed, the ten Commandments, and about half a dozen verses besides of the Bible*, whose *lines* were so *small* and *near together*, that I was unable to *number* them with my *naked eye*".[57] Both metaphorically and literally, in *Micrographia* we witness Hooke learning to read and write in a new way, through the magnifying lens, taking as his first texts precisely the religious reading matter often found alongside the alphabet in children's hornbooks and primers.[58]

Hooke, moreover, did not simply intend for *Micrographia* to illustrate what he had observed through the microscope; he also wanted it to recreate, for the reader, the process and experience of such observation.[59] Meghan Doherty, for instance, has shown how Hooke uses representational techniques including the enclosure of "the object of study … within a perfect circle" to create an illusion of looking through the microscope's eyepiece. "What the viewer is presented", Doherty comments, "is not what was put on the stage of the microscope, but rather what resulted from the act of looking through the microscope".[60] Perhaps, too, the sheer size of *Micrographia*—a folio made even larger by the folded pages the reader must unfurl, including the famous four-page engravings of a louse and flea—can be understood in this context. It may, that is, be understood not only as a bid for visual impact, but as part of Hooke's effort to recreate the experience of using the microscope—an experience, I have suggested here, that was characterized by a sense of temporal regression, a partial restoration of youth's sensory privileges and scholastic duties. Opening Hooke's book to find familiar things enlarged sometimes

---

[57] Hooke, *Micrographia*, B1r (1), B3r (3).

[58] On the inclusion of religious materials in hornbooks and primers, see Lamb, *Reading Children*, 29–31. Cowley, too, describes the experience of looking through a microscope as akin to the experience of a child who has just begun to be able to decipher and interpret a difficult text. "Y' have learn'd to Read her [nature's] smallest Hand", he congratulates in his commendatory poem on the Royal Society, "And well begun her deepest Sense to Understand". Cowley, "To the Royal Society", B3r (unpaginated).

[59] Hooke thereby participated in the broader project of developing what Shapin and Schaffer, with a focus on Boyle, influentially described as technologies of "virtual witnessing", that is "the production in a *reader's* mind of such an image of an experimental scene as obviates the necessity for either direct witness or replication". Shapin and Schaffer, *Leviathan and the Air-Pump*, 60.

[60] Doherty, "Discovering the 'True Form'", 215.

past recognition may produce a befuddling sense of being a child operating in a world which is simply too big; turning its pages and unfolding its illustrations, perhaps we are meant to feel in some sense infantalized, carried back to a time when picture-books and primers still felt unwieldy in our hands.[61]

*   *   *

As Sorana Corneanu has shown, seventeenth-century experimental science was a form of *cultura animi*, intended not only to provide knowledge of the natural world but also to inculcate virtue, curing or reforming the experimental philosopher's own errant mind.[62] In Thomas Sprat's words, "*Experiments* ... are able to give [assistance] towards the management of the privat *motions*, and *passions* of our *minds* ... the *Real Philosophy* will supply our thoughts with excellent *Medicines*, against their own *Extravagances*, and will serve in some sort, for the same ends, which the *Moral* [Philosophy] professes to accomplish".[63] In this chapter, I have suggested that this project of entwined epistemological and ethical reformation can also be understood, in part, as a process of regression to a childlike state, in which experiment is both a form of innocent play and a means of restoring the perceptual clarity of youth. Experiment may often be reliant on technological sophistication, but it also privileges a form of experience which is momentary and exceptional, rather than cumulative and quotidian—and hence, accessible to children as well as adults. Hooke's *Micrographia*, in particular, is intended partly to recreate the experience of seeing as a child sees, as the toy-like technologies of the microscope and telescope allow Hooke to recapture something of the immediacy, the acuity, and the wonder of juvenile perception.

What, though, of the flipside of this conceptual association? If experiment was considered a kind of childish play, might childish play, conversely, be considered a form of experiment? Modern theorists often do

---

[61] Matthew C. Hunter comments: "Daunting a 'textbook' as it is, minor circumstantial evidence suggests that segments of Robert Hooke's audience actually did try to learn microscopic seeing and drawing from works like *Micrographia*". Most notably here, "a young Isaac Newton made drawings from *Micrographia*". Hunter, *Wicked Intelligence*, 43.

[62] Corneanu, *Regimens of the Mind*.

[63] Sprat, *History of the Royal Society*, Uu3v (341).

think of it in these terms: in the twentieth and twenty-first centuries, a range of psychologists, philosophers, and educational theorists have argued that children are, as the influential child development researcher Jean Piaget postulates, "little scientists".[64] As Huizinga notes in *Homo Ludens*—citing Piaget—"experimental child-psychology has shown that a large part of the questions put by a six-year-old are actually of a cosmogonic nature, as for instance: What make water run? Where does the wind come from? What is dead?"[65] More recently, Alison Gopnik has elaborated on Piaget's suggestion, arguing in her book *The Philosophical Baby* that infants and young children acquire knowledge (including causal knowledge about their physical environment) using experimental processes similar to those employed by scientists.[66]

Such work, however, tends to limit the utility of the experimental "data" produced by children to those children themselves—infants only learn what adults already understand. In contrast, I will demonstrate in the next chapter of this book, early experimental philosophers not only recognized analogies between their own developing methodological commitments and the habitual activities of children; they also took child's play as an illuminating source of new knowledge about the world, conceptualizing it as a form of experiment which was productive precisely insofar as it was purposeless.

---

[64] See Kuhn, "Piaget's Child as Scientist".

[65] Huizinga, *Homo Ludens*, 107; See also Piaget, *The Language and Thought of the Child*, Chapter 5.

[66] Gopnik, *The Philosophical Baby*, especially 74–95; see also Gopnik, Meltzoff, and Kuhl, *The Scientist in the Crib*. Gopnik also takes issues with many aspects of Piaget's work, such as his contention that young children cannot distinguish between fantasy and reality. On the philosophical capacities of children more broadly, see Matthews, *Philosophy and the Young Child* and *The Philosophy of Childhood*.

CHAPTER 5

Bubbles, Popguns, Lizards' Tails: Play as Experiment

Abstract This chapter argues that if, in the seventeenth century, experiment was thought of a form of childish play, the converse was also true: play was thought of a form of experiment. The pervasive framing of the natural sciences as play in the work of seventeenth-century experimental philosophers, moreover, is not merely rhetorical hot air: actual, real-life children were active participants in the natural philosophical world of seventeenth-century England, both in semi-professional roles, as assistants and apprentices, and in more informal, domestic settings. In particular, children's games were treated as a kind of spontaneous, generatively erratic, imaginatively fertile, ad hoc experimentation, from which quite sophisticated theories about the natural world might eventually develop. Notably, bubble-blowing played a critical role in the development both of Francis Bacon's ideas about the nature of fluids and of Isaac Newton's groundbreaking theories of light and colour. The final part of the chapter discusses the presence of *putti* in contemporary scientific illustration, suggesting these indicate the varied roles played by children in the early modern natural and experimental sciences.

Keywords Children • Science • Natural philosophy • Experiment • Royal Society • Play • Tycho Brahe • Robert Hooke • Robert Boyle • Francis Bacon • John Beale • John Ray • Francis Willughby • Bubbles • Isaac Newton • Rene Descartes • Bernard Le Bovier Fontenelle • *Putti* • Scientific illustration • Maria Sybilla Merian

© The Author(s), under exclusive license to Springer Nature
Switzerland AG 2024
E. L. Swann, *Science as Child's Play in Seventeenth-Century England*, https://doi.org/10.1007/978-3-031-75849-2_5

In his *Certain Physiological Essays* (1661), Boyle explains why, in the work, "I have not refrain'd from mentioning divers things which may seem very slight, because very obvious". His reasons are both pragmatic—simpler experiments are "more easily and cheaply try'd"—and principled, originating from the conviction that even apparently insignificant activities can be illuminating for the natural philosopher. "I disdain not", he proclaims, "to take Notice ev'n of Ludicrous Experiments, and think that the Plays of Boys may sometimes deserve to be the Study of Philosophers".[1] Here, Boyle aligns "Experiments" and boys' games: both are "Ludicrous"—a word which derives from the Latin *lūděre*, to play—in a sense which simultaneously admits and dismisses the derisive quality of modern uses of the term, invoking instead the word's now-lost, non-pejorative, etymological meaning of simple playfulness.

Anticipating that opponents of the Royal Society will object to its impact on education, in the *History of the Royal Society* (1667) Sprat claims that "Men are not ingag'd in these *Studies*, till the Course of *Education* be fully completed ... the *Art* of *Experiments*, is not thrust into the hands of Boyes, or set up to be perform'd by Beginners in the School".[2] Whatever its success or otherwise in reassuring the Society's detractors, however, Sprat's assertion that children have no part to play in "the Art of Experiments" is belied by the evidence.[3] We have already seen, in Chap. 2, that Beale drew on the supposed acuity of children's senses in his investigations of luminescence. As this chapter will show, this was not an isolated incident: children were active contributors to the development of early modern natural philosophy, including the experimental world of the early Royal Society—if not at "School", certainly in a range of domestic, social, and semi-professional contexts.

Perhaps most obviously, older children and adolescents had a formal role, as students, assistants, and apprentices. Figure 5.1, an engraving from a hand-coloured presentation copy of the *Astronomiae Instauratae Mechanica* [*Instruments of the Renewed Astronomy*] (1598) by the brilliant Dutch astronomer Tycho Brahe—himself allegedly "Proficient" in

---

[1] Boyle, *Certain Physiological Essays*, D1r (17).

[2] Sprat, *History of the Royal Society*, Ss2r (323). In fact, in 1661 Abraham Cowley had proposed the founding of "a *Philosophical Colledge*" incorporating a school of around two hundred boys who were to receive an "Apprenticeship in Natural Philosophy"; this plan did not come to fruition. Cowley, *A Proposition*, A5r, A7r, C7r.

[3] Notoriously, Sprat's *History* was a flop as propaganda; see Hunter, *Establishing the New Science*, 63–64.

Fig. 5.1 Mural quadrant, from Tycho Brahe, *Astronomiæ Instauratæ Mechanica* (Wandesburgi, 1598). © Bodleian Libraries, University of Oxford. Shelfmark Arch. B c.3. Terms of use: CC-BY-NC 4.0

studying the heavens "even from his Childhood"—depicts in the fore-ground three figures in the process of making observations with Brahe's famous mural quadrant (an astronomical instrument used to calculate the altitude of the stars).[4] Constructed in 1582, the quadrant—the curve of which bisects the image—consisted of a solid brass arc with a radius of six-and-a-half feet, affixed to a wall of Brahe's observatory on the island of Hven, the quarter circle circumference of which Brahe took some trouble to have decorated with a mural.[5] Depicting Brahe himself as the seated, grandly gesticulating figure who dominates the engraving, the mural also shows the three main locations of his work, just visible through the arches in the background: at the bottom, his alchemical laboratory; in the mid-dle, his study; and at the top, an outdoor space populated with some of Brahe's other astrological instruments.[6]

As John Robert Christianson has documented, Brahe's research relied heavily on an extensive network of assistants and collaborators drawn from a motley mix of social groups, including students (often from the University of Copenhagen), artisans, servants, local peasants, and Brahe's own rela-tives.[7] This group, which he described as his *familia*, comprised individu-als of a wide range of ages, including children and adolescents; Tycho's observations of a lunar eclipse in 1573, for instance, were aided by his fourteen-year-old sister, Sophie Brahe, and admission to the University of Copenhagen could take place as early as fourteen years of age.[8] Figure 5.1 offers a striking representation of this diversity: thus, Brahe's study, as he explains in his explanation of the engraving, includes "tables ... at which my assistants (who numbered six or eight, sometimes ten or twelve, in

---

[4] The copy in question was presented to Marino Grimani, the Doge of Venice, in 1599, and now resides in the Bodleian Library, Oxford; see https://digital.bodleian.ox.ac.uk/obj ects/0679de02-6818-4885-8c2e-77df05dbe3ab/. On the young Brahe's enthusiasm for astronomy, see Sherburne, *The Sphere of Marcus Manilius*, Q1r. (63). Sherburne writes that "even from his Childhood being addicted to Astronomical Studies ... [Brahe] grew by his own Ingenuity and Industry without any Instructor, so great a Proficient therein, that in the time of his Minority, and without the help of other Instruments, than a small Globe little bigger than a Man's Fist, and a large pair of Compasses, with which ... he used by stealth to take the Distances of the Stars, he made a shift to detect divers considerable Errors, both in the *Alphonsine* and *Prutenick* Tables".

[5] For discussion of this engraving, see Horacek, "Illuminating Methods", 32, 52–53.

[6] Brahe, *Instruments of the Renewed Astronomy*, 30–35.

[7] Christianson, *On Tycho's Island*, 58–9.

[8] Christianson, *On Tycho's Island*, 57.

addition to boys and young students) could sit when making calculations".[9] The "collaborator at H", too—depicted facing away from the viewer in the foreground of the image—who "watches the clocks", ensuring the accurate timing of the observations made by the figure the far right of the image (labelled F), appears both by his physique and by his close-cropped hair to be a boy.[10]

In England in the following century, Hooke, too, was not unusual in taking boys (who often lodged with him) as apprentices and assistants.[11] Perhaps the most talented of these youngsters was Harry Hunt, who—as the editors of Hooke's diary note—joined his household "as a boy" in January 1673, adopting the positions both of student or mentee, and of employee or junior colleague.[12] In the weeks and months following his arrival, Hooke's diary tracks Hunt's increasing abilities. On 27 January that year, Hooke records that "with Harry I wrought on the Specular metall for Telescope"; on 25 February, he was with "Harry about fire Experiment with funnell"; on 1 and 25 April, he "viewd ♀ [Venus] and ♃ [Jupiter] with Harry".[13] As the year progressed, Hooke seems to have entrusted Hunt with more independent work: in September, he reports that "Harry made microscope tooles of 2/5 of an inch".[14] A few years later, in 1676—and, it seems, somewhat to Hooke's surprise—Hunt was given official employment as the Royal Society's "Operator", responsible for any number of practical or mechanical tasks that might be assigned to him.[15]

Barred from apprenticeships and positions of professional responsibility, some girls, too, nonetheless pursued instruction in the new sciences from male relatives, friends, and acquaintances, as Patricia Fara has shown. "Even as a child", for example, Catherina Koopman—later Elisabetha Hevelius, wife to Johannes Hevelius, and a brilliant astronomer in her own right—"had wanted to study the stars". She achieved her aim at the age of sixteen, Fara suggests, by marrying Hevelius, "reminding him of a childhood promise he had made to teach her astronomy".[16] Similarly, Maria

---

[9] Brahe, *Instruments of the Renewed Astronomy*, 30–35.
[10] Brahe, *Instruments of the Renewed Astronomy*, 30–35.
[11] Henderson, "Introduction", 132.
[12] Hooke, *The Diary*, 498; see also Jardine, *The Curious Life*, 315.
[13] Hooke, *The Diary*, 24–5.
[14] Hooke, *Diary*, 60.
[15] Hooke, *Diary*, 255. On Hooke's training of Hunt, see Espinasse, *Robert Hooke*, 131–2.
[16] Fara, *Pandora's Breeches*, 130–31.

Winkelmann (later Maria Kirch) "was interested in astronomy as a child"; later, she "taught practical astronomy to her own children".[17] In England, Robert Hooke's teenage niece Grace, who first joined his household in around 1670, when she would have been ten years old, received instruction in algebra from her uncle, as well as borrowing books including Peter Heylyn's *Cosmographie* (1652) from him.[18] The young Anne Conway, meanwhile, "begged her brother to keep her up-to-date with what he was learning at Cambridge".[19] In this, Conway was probably not alone: a number of the popularizing guides to natural philosophy that began to appear in the late seventeenth century—themselves intended for a readership both of "young Gentlemen" and of "young ... Ladies"—took the form of dialogues between an older brother, recently returned from university, and a younger sister with whom he shared his learning: a literary convention which, as the example of Conway suggests, may also reflect the experiences of readers.[20] By the early eighteenth century, the figure of the "Philosophical Girl" was well-established enough to be (sympathetically) lampooned: in Susannah Centlivre's anonymously published comedy *The Basset-Table* (1705), performed at London's Drury Lane Theatre in 1706, the "little she-Philosopher" Valeria reads Descartes, anatomizes dogs, and practices microscopy in order to witness the circulation of blood.[21]

Fusing the traditional role of apprentice with that of laboratory technician, the young Harry Hunt worked for Hooke in a semi-official capacity. Often, however, children participated in experimental activities in a more informal, playful, ad hoc way—something that should not, perhaps, be surprising, given that such activities were not limited to the rarefied space of the laboratory in this period, but took place in range of quotidian domestic contexts, including kitchens, bedrooms, and gardens.[22] Here, as

---

[17] Fara, *Pandora's Breeches*, 140–41.

[18] On 30 June 1676, Hooke records that he "began first to read algebra to Grace and Thom"; on 03 July, he again "read algebra to Thom and Grace". On 9 January 1676/7716, "I lent Grace, Hylens Cosmography and Maps". Hooke, *Diary*, 239–240, 267. Hooke's concern for the education of his young charge was presumably not purely altruistic: beginning on 04 June 1676, not long after she turned sixteen and just before he begins to mention her intellectual interests, he began to have sex with her.

[19] Fara, *Pandora's Breeches*, 111.

[20] See, for example, Martin, *Young Gentlemen's and Ladies Philosophy*, discussed in Fara, *Pandora's Breeches*, 109–10. On "the gendered character of eighteenth-century natural history instruction", see Stafford, *Artful Science*, 63–67 (quote at 64).

[21] Centlivre, *The Basset-table*, 19. On the play, see Fara, "Elizabeth Tollet", 174.

[22] See Shapin, "House of Experiment".

many scholars have shown, adult female family members and domestic workers also contributed actively and substantially to the production of knowledge.[23] Early modern England was a notably young society, with individuals under the age of fifteen making up over 30 per cent of the population by the end of the seventeenth century.[24] Children, then—both girls and boys—would also inevitably have been present in these spaces. Notably, as Julie Davies has shown, Mary Somerset, first Duchess of Beaufort, was at various points assisted in her botanical work by her children, especially her eldest son Charles, who "held a particular interest in his mother's pursuits" and who, "in 1673, at the tender age of 13, was the youngest nominee to be elected a Fellow of the Royal Society, a distinction he still holds".[25]

In Chap. 2, I discussed John Beale's attempt to determine variations in the temperature of some pieces of shining pork. This was not, however, the first time that Beale had encountered luminescence in his larder, nor was it the first time that he had benefited from the assistance of the younger members of the household. In a letter published in the *Philosophical Transactions* over a decade earlier, in 1665, Beale describes how he was alerted by his cook to some pickled mackerel which had begun to emit a strange radiance. "The Cook stirring the water, to take out some of the Mackrels," he records, "found the water at the first motion become very luminous, and the Fish shining through the water". He goes on to detail the behaviour of this water when it is dispersed:

> Wherever the drops of this water (after it was stirr'd) fell on the ground, or benches, they shin'd: And the children took drops in their hands, as broad as a penny, running with them about the house, and each drop, both neer and at distance, seemed by their shining as broad as a six-pence. … On Tuesday night (May 9) we repeated the same Trial and found the same effects. … In these Vulgarities we may perhaps as well trace out the cause and nature of Light as in Jewels of greatest value.[26]

---

[23] See, *inter alia*, DiMeo, *Lady Ranelagh*; Hunter and Hutton, eds., *Women, Science and Medicine*; Zinsser, ed., *Men, Women, and the Birthing of Modern Science*; Hutton, "Science and Natural Philosophy"; Hutton, "Alchemy and Cultures of Knowledge"; Leong, *Recipes and Everyday Knowledge*; Long, ed., *Gender and Scientific Discourse*; and Wall, *Recipes for Thought*.

[24] Thomas, "Children", 51.

[25] His participation in the Society's activities, however, was fairly minimal. Davies, "Botanizing at Badminton House", 22, 30.

[26] Beale, "An Experiment", 1226–228.

In his investigations into the pork, as discussed in Chap. 2, the involvement of the children is actively sought and stage-managed by Beale: "I desired all the company, (whereof some were young children, which have the tenderest touch) to try, whether the most flaming parts had any perceptible degree of tepidity".[27] By contrast, here we get a sense of the children's involvement as a spontaneous and autonomous response to this strange, delightful phenomenon. The unpremeditated, playful quality of their reaction does not, however, reduce its utility: their excited dispersal of the shining "drops" is subsequently framed by Beale as a "Trial" or experiment, which he finds interesting enough to repeat, with identical effects. This "Trial", moreover, has far-reaching implications: the enduring brightness of the scattered water is valuable information, which might eventually contribute to nothing less than a better understanding of "the cause and nature of Light" itself.

It is unclear whether the children Beale describes were aware of any broader significance their actions might have. Elsewhere, however, the contributions made by children to early experimental philosophy were very obviously unintentional. In the seventeenth century as today, childish curiosity and impetuosity—not to mention their tendency, noted by Francis Willughby, "to put every thing into their mouths"—resulted in a predisposition to accidents, the consequences of which are observed in some early issues of the *Philosophical Transactions*.[28] In "An Account of the Plant, call'd Bangue"—that is, marijuana—delivered to the Royal Society in December 1689, for instance, Hooke notes that "I have formerly given an Account of the Effects of the Roots of Hemlock, accidentally eaten by some young Children, which, at first, had an Operation on them much of the like Nature with this Vegetable [i.e. bangue]".[29] Here, the childish propensity to ingest dangerous substances—and its narcotic

---

[27] Beale, "Two Instances of Something Remarkable", 601.

[28] Willughby, *Book of Games*, 187.

[29] Hooke, "An Account of the Plant, call'd Bangue", 210. Hooke goes on to describe the effects of ingesting "Bangue" as follows: it "doth, in a short Time, quite take away the Memory and Understanding; so that the Patient ... in that Extasie ... becomes, as it were, a mere Natural, being unable to speak a Word of Sense yet is he very merry, and laughs, and sings ... after a little Time he falls asleep ... and when he wakes, he finds himself mightily refresh'd, and exceeding hungry". On Hooke's own frequent ingestion of a range of intoxicants and narcotics, see Jardine, *The Curious Life*, 280–81; and Brown, Plunkett, and Yates, "Hooked".

or intoxicating effects in the case of hemlock—is marshalled as part of Hooke's attempt to taxonomize and make sense of the unfamiliar cannabis plant.[30] In this regard, the impulsiveness and volatility of children introduced into early modern natural philosophy an element of the generatively random, accidental quality which Michael Witmore identifies as a key component of the Baconian experiment itself.[31]

Elsewhere, the unpredictable quality of child's play has effects which in the immediate term are somewhat violent, but which in the longer term are seen to have the potential to lead to significant medical advances. In his *Observations Touching the Usefulness of Experimental Natural Philosophy*, Boyle argues that "the Doctrine so unanimously delivered by Physitians and Chirurgions, concerning the irreparable loss of the Limb of an Animal, once violently severed from the Body", is "unfit to be admitted", noting the counterexample of lizards, whose "Tails being struck off will grow again". This, he states, "hath been of old observ'd by *Pliny*".[32] Ancient authority, however, is bolstered by the more recent testimony of the Dutch physician and naturalist Jacob de Bondt, who reports that:

This I have more then once Observ'd in *Lizards* which I kept in my own House. For my Children being at play, when with a Rod they had strook off the *Lizards* Tails I saw them within a day or two come out to Feed, and their Tayles then by little and little still encreasing and growing bigger.[33]

---

[30] For another example of the childish propensity to ingest inappropriate substances (in this case, "two copper farthings") marshalled as natural history, see Underhill, "An Account", 424.

[31] Witmore argues that "there are at least two fundamental similarities between accident and experiment in Bacon's natural philosophy: both are a result of an unusual disposition of circumstances, and both are understood to be a more 'artificial' version of what nature does when left to its own habits". Accidents "also represented an instance of deviation in the usual course of things; their power to distract or estrange onlookers from habitual patterns of expectation and attention thus gave them unusual epistemological force". Witmore, *Culture of Accidents*, 126.

[32] Boyle, *Observations Touching the Usefulness*, clv (18). For Pliny's comments on this phenomenon, see Pliny *Natural History*, 221. One of my own earliest memories is being told that if a worm was cut in half, the individual pieces would become two separate worms. I secretly decided to try this, but—immediately horrified by what I'd done—hid both parts in a compartment under the seat of my plastic tricycle, where they presumably withered and died (I wouldn't go near the tricycle for months after). Perhaps this is why I'm a humanist, not a scientist.

[33] Boyle, *Observations Touching the Usefulness*, Ddd3r (405); original Latin at C2r (19).

Here, the impulsive cruelty of children at play ultimately offers some veri-fication for the possibility that limbs, once severed, may regenerate. In so doing, it offers hope to those whose ambitions for the new experimental philosophy included the possibility of healing abilities beyond the dreams of conventional seventeenth-century medicine.

Childish vulgarity, too, has a place in the annals of early modern natural philosophy. In his 1673 travelogue, *Observations Topographical, Moral, & Physiological,* John Ray describes an encounter with a Venetian nobleman who:

> shewed us a Boy, who had a faculty of charging his belly with wind, and discharging it again back ward at pleasure; which we saw him perform. When he charged himself he lay upon his hands and knees, and put his head on the ground almost between his legs. The same Nobleman shewed us the experiment, and gave us the receipt of a fulminating powder [i.e. a combus-tible powder].[34]

Amusingly crude anecdote this may be, but the boy's ability to pass wind at will, aided by his gymnastic contortions, is consequential enough to be accorded the name of "experiment" by Ray, who sees fit to record it along-side his acceptance of a different kind of controlled explosive: a recipe for "fulminating power".

Ray was accompanied on his travels by his close friend and colleague, the botanist, entomologist, ichthyologist, ornithologist, astronomer, and Fellow of the Royal Society Francis Willughby: a man who, as the main author of the manuscript known as the *Book of Games,* also gave child's play some serious consideration. Unfinished at Willughby's death, aged thirty-six, in 1672, the *Book of Games* is a descriptive compilation of a wide variety of English "plaies" (including both children's games and more adult-oriented pastimes) which registers Willughby's scientific interests in a number of ways.[35] Most obviously, its taxonomic structure reflects con-temporary efforts (including by Willughby himself, as well as Ray) to

---

[34] Ray, *Observations,* O5v (202). On Ray and Willughby's travels, and the resulting *Observations* as "a model for how to 'travel scientifically'", see Greengrass *et al.,* "Science on the Move", 152–197 (quote at 197).

[35] Willughby, *Book of Games,* 93–4. On Willughby's membership of the Royal Society, see Cram, Forgeng, and Johnston, "Introduction", 16–22; on "the Baconian tradition" as a pertinent context for the work, see "Introduction", 52–59; on other "mathematical and scientific aspects", see "Introduction", 60–64.

develop classificatory schemes across a range of fields.[36] Willughby's mathematical interests, too, are evident in his numerical analyses of dice and card games, as well as the scoring patterns of certain kinds of athletic games (including tennis and bowls), which he connects to astronomical calculations.[37]

While the complexity of such games meant they were largely associated with adults, Willughby's descriptions of specific children's games such as "Copshole" (whereby "2 boyes sitting downe upon the ground set the soles of their shooes one against another, & holding a stick in their hands pull as hard as they can"), "Pitching the Barre" (a throwing game involving "a great long barre of iron"), and "Running Jump" (a form of competitive leaping) are often attentive to their underlying physics, especially motion—a topic Willughby also explored in a series of 1669–1670 letters to Henry Oldenburg. Here, Willughby criticized the theories of motion associated with Christopher Wren and Christiaan Huygens on the basis that they lacked experimental proof.[38] As David Cram, Jeffrey Forgeng, and Dorothy Johnston suggest in the "Introduction" to their recent edition of *Book of Games*, then, for Willughby "games"—including children's games—provided something like "a natural laboratory", where the forces of nature could be observed quite literally at play in the interactions between human bodies and physical objects.[39]

Willughby's eclectic range of interests included worms, comets, coal, spiders, coins, the moon, and walnuts, and his gathering "of Birds, Fishes, Shells, stones and other fossils, seeds, dried plants, coins etc." (as described by Ray in a letter to Martin Lister) place him firmly in the camp of collectors mocked as childishly indiscriminate by Judith Drake in Chap. 3.[40] In the *Book of Games*, however, he also enlisted the help of an anonymous young collaborator: as Cram, Forgeng, and Johnston note, features of the

---

[36] As Cram, Forgeng, and Johnston write, Willughby's classifications "show him applying the same techniques of scientific description to games and recreations that he was employing in his natural history work". Cram, Forgeng, and Johnstone, "Introduction", 40; on the "striking parallels" between Willughby's taxonomic scheme and that of John Wilkins in his *An Essay towards a Real Character, and a Philosophical Language* (1668), see "Introduction", 54–6.

[37] See Cram, Forgeng, and Johnston, "Introduction", 60–64.

[38] Willughby, *Book of Games*, 170–72. On the letters and their reception, see Cram, Forgeng, and Johnston, "Introduction", 19.

[39] Cram, Forgeng, and Johnston, "Introduction", 63.

[40] Ray to Lister, 18 June 1667, in Ray, *Further Correspondence*, 112.

manuscript (including the ruling of the pages in question, the use of pho-
netic spelling, and corrections made by Willughby himself) indicate that
the writer responsible for the section on children's "plaies", as well as a
few other juvenile games scattered throughout, "was somebody with
imperfect writing skills, and most probably a child".[41] In treating children's games as an object of scientific scrutiny and a source of scientific
knowledge, then, Willughby did not rely solely on his own memories or
observations; instead, he enlisted the expertise of an associate with more
recent direct experience of the topic at hand.

It is possible that Willughby began the *Book of Games* in response to
Francis Bacon's call, in the *Novum Organum*, for a "history of man"
which would include within its scope a "Historia Ludorum omni generis
[History of games of all kinds]".[42] In fact, Bacon himself may at one point
have intended to begin such a "History". Amongst his loose papers, there
is sheet of notes in Bacon's own handwriting on the subject of "Play",
gesturing to the moral and social dimensions of games (which, he notes,
may be "cause of oths, quarells, expence and … ydlenes", but which are
also sources of "society, acquaintance, familiarity in frends … recreatio[n]
and putting of melancholy") as well as possible taxonomical distinctions
(he distinguishes, for instance, between games of "hazard" and "cun-
nyng"; "frank" or "wary" games; and "quick" or "slowe" ones).[43] And
although Willughby didn't manage to complete his collection of "plaies"
before his early death, he and his juvenile assistant nonetheless made more
progress than their predecessor: Bacon's own brief notes on the topic are
rough jottings, written in haste, which never—as far as we know—pro-
ceeded past a very preliminary note-taking stage.

Despite this, Bacon's interest in the value and importance of child's
play—particularly for understanding the nature and motion of matter—is
evident in several places throughout his work. In particular, we can see
children's games functioning as a kind of experiment in the *Novum*

[41] The identity of the child in question is not specified in the text; see Cram, Forgeng, and Johnston, "Introduction", 42.

[42] Bacon, *Novum Organum*, 11.479, 11.484; on the possibility that Willughby's work was "a response to Bacon's call", see Cram, Forgeng, and Johnston, "Introduction", 54.

[43] Bacon, notes on "Play", in *OFB*, 1.568. On the possibility that these notes represent an early, abandoned effect to write a *Historia Ludorum omni generis*, see the editorial notes to Bacon, *The Works*, 7.210. On this document, see Moshenska, *Iconoclasm as Child's Play*, 179–80.

Organum itself, where Bacon discusses a specific class of phenomena he calls "*Clandestine Instances*": things which "show the nature under investigation with its virtue at its lowest". Clandestine instances, that is, are things that display a specific characteristic or nature, but only weakly or faintly—as Bacon says, "as if the nature were in its cradle and rudiments". Such "instances", Bacon argues, "are of the very first importance for discovering forms", for they "best lead to genera, i.e. to those common natures of which the natures under investigation are nothing other than limitations". For example:

> Let the nature under investigation be consistent or determinate, the opposite of which is liquid or flowing. *Clandestine Instances* are the ones which manifest some slight of very low degree of consistency in a fluid; as a bubble of water which is like a kind of consistent and determinate membrane made from the water's body ... like this is the instance of children's mirrors, which the little ones usually make from saliva on rushes, where we also see a consistent membrane of water. But much better does this show itself in that other children's pastime, i.e. when they take water, make it a little more clinging with soap, blow it through a hollow reed, and so turn the water into a kind of bubble castle which, by the inclusion of air, becomes consistent enough to stand being thrown some distance without loss of continuity. ... All of this clearly suggests that liquid and consistent are only vulgar notions based on our senses; and yet that all bodies really do have a tendency to fly and avoid loss of continuity, a tendency which in homogeneous bodies (such as liquids) is weak and feeble, but in heterogeneous ones stronger and more forceful.[44]

Bacon uses the example of children spitting on rushes to make imitation "mirrors" and blowing clusters of soap bubbles which have enough tenacity to be thrown short distances in order to put forward the physical principle that "consistent and fluid" are not absolute, but rather relative, states. All bodies have a tendency to some kind of consistency or form, and this is simply weaker in homogenous bodies, and stronger in heterogeneous ones.

In the same work, Bacon describes a form of motion whereby "bodies exert themselves to be free of preternatural pressure or stretching". "Examples of this motion", he continues, "are countless":

---

[44] Bacon, *Novum Organum*, 11.281–83.

... as, for instance (as far as liberation from pressure is concerned), of water in swimming, air in flying, of water in rowing, air in the gusting of winds, and of springs in clocks. Nor does the motion of compressed air show itself ungracefully in children's pop-guns [*in Sclopettis ludicris puerorum*], when they hollow out alder or something like it and block it at both ends with a bit of some sappy root or other, then with a ramrod in the neat end they force the root or stuffing out the fat one—as a result of which the root is sent or shot out of the far end with a bang, and that before the ramrod or stuffing in the neat end has touched it.[45]

In both cases, Bacon presents the children's games he describes as "examples" or "instance[s]" of a physical principle; at first blush, they are merely a means to communicate, in an immediately recognisable way, a theory or natural process. Despite Bacon's own rhetorical framing, however, these examples do have evidentiary, as well as merely illustrative force: the bubbles example is presented as corroboration for the assertion that "that liquid and consistent are only vulgar notions". Bacon is not simply using the example of children's games to illustrate a physical principle which is already established. Rather, he is using his close, attentive observations of those games both to develop, and to demonstrate the legitimacy of, those principles: to create knowledge, not just to communicate it. The games themselves function as experiments, performed by children, which Bacon witnesses and learns from. Perhaps, then, Bacon's presentation of them as "examples" is not a genuine reflection of his attitude to their epistemic potential, but rather a rhetorical strategy intended to suggest that the validity of the physical principles and processes he describes is a *fait accompli* which requires explanation but not verification.[46]

[45] Bacon, *Novum Organum*, 11.385; see 11.384 for the Latin. See also Bacon, *Phenomena of the Universe* [*Phaenomena Universi*] in *OFB*, 6.38–9: "In an effort to avoid pressure, air produces and mimics all the effects of a solid and robust body ... Children hollow out alder wood in imitation of guns, and stuff pieces of iris root or paper pellets into each end of the tube, and then thrust out the pellet by pushing it with a wooden ramrod, but the pellet at the other end is discharged with a violent bang by the force of the enclosed and compressed air before the ramrod touches it at all". For this latter reference, I am indebted to Sophie Weeks.

[46] See also the *Sylva Sylvarum*, where Bacon notes that "all *Eruptions* of *Air*, though small and slight, give an *Entity of Sound*. ... As in ... *Green Wood* laid upon the fire ... so in *Candles* that spit Flame, if they be wet ... So in a *Rose leaf* gathered together into the fashion of a Purse, and broken upon the Forehead, or Back of the Hand, as Children use". Bacon, *Sylva Sylvarum*, D5v (34).

Also significant here is the way that the fanciful character of the child-ish, playful manipulations of the natural world that he describes helps pro-pel Bacon and the reader alike beyond their own erroneous assumptions. As scholars including Mary Baine Campbell, Elizabeth Spiller, and Frédérique Aït-Touati have shown, for all of Bacon's vociferous suspicion of the wayward faculty of the fancy, imaginative speculation had a range of vital functions within early modern natural philosophy, serving—*inter alia*—as an impetus for speculative hypotheses, as a creative force in the construction of experiments, and as a means of visualizing distant or com-plex phenomena that could not otherwise be directly observed.[47] Here, the children's activities effect material transformations that reveal underly-ing physical principles. But they do so, crucially, by simultaneously engag-ing the imagination both of the children involved and of the natural philosopher who watches them, as a means of breaking down quotidian, common-sense, but misguided assumptions.[48] In this make-believe world where "mirrors" are actually spit, where foamy "castles" rise and fall, and guns can be formed of alder twigs, physical principles such as "liquid and consistent", which may seem self-evident, are also revealed as contingent fictions: imaginative play reveals that they are "only vulgar notions based on our senses", with little connection to underlying realities.

Although the scientific significance of pop guns is not a prominent theme in the history of physics, Bacon's interest in bubbles is worth dilat-ing on: in later seventeenth-century philosophy, soap bubbles often pop up as key to the development of theories of light and colour, from the mechanistic models proposed by Boyle and Hooke, to Newton's theory that colours are intrinsic properties of light itself.[49] In his *Experiments and Considerations Touching Colours* (1664), for example, Boyle refutes the

---

[47] Campbell, *Wonder and Science*; Spiller, *Science, Reading, and Renaissance Literature*; Aït-Touati, *Fictions of the Cosmos*. See also Raymo, "Science as Play", 284–5.

[48] In his essay "Of Youth and Age", Bacon remarks on the imaginative vitality of youth (in this case, young adults): "the Invention of *Young Men*, is more lively, then that of Old: And Imaginations streame into their Mindes better". Nonetheless, "generally, *youth* is like the first Cogitations, not so Wise as the Second". Bacon, *The Essayes*, Ii4r (247). The first (1612) edition of this work does not include the comment about "the Invention of *Young Men*".

[49] On the role of bubbles in the history of optics and physics, see Schaffer, "A Science whose Business is Bursting". Sarah Tindal Kareem comments that "Hooke's and Newton's observations of soap-bubbles affirm the presence of the profound within apparently 'childish and insignificant' pursuits", as part of her broader argument that eighteenth-century analo-gies between bubbles and poetry draw on experimentalism in order to "articulate a new idea of literature that defines the act of literary composition as a form of play that brings forth new worlds". Kareem, "Enlightenment Bubbles", 98, 85.

chemical theory of colours, according to which chromatic variations are produced by different combinations of the hypostatic principles of salt, sulphur, and mercury, by urging "the Chymists" to consider "that some [Colourless] Bodies ... by being brought to a great Thinness of parts, acquire Colours though they had none before", giving as an example "the Variety of Colours, that Water, made somewhat Glutinous by Sope, acquires, when 'tis blown into such Sphaerical Bubbles as Boys are wont to make and play with".[50] The observation that the water manifests colour not when the soap is added, but when it is blown (by "Boys") into "Sphaerical Bubbles", is levied as evidence for Boyle's corpuscularian theory that colour arises from minute textural alterations in the surface of a body, rather than from the chemical constitution of that body. Similarly, in his *Micrographia*, Hooke notes that he has "often observed" the "Colours" visible in "those Bubbles which Children use to make with Soap-water", taking this phenomenon—alongside his microscopical observations "Of the Colours observable in Muscovy Glass" (i.e., the mineral today known as mica or isinglass)—as evidence that that "the material cause of the *apparition* of these several Colours" is "the greater or less refractive power of the *pellucid* body".[51]

Most consequentially for the subsequent history of physics, Isaac Newton made bubbles central to his *Opticks* (1704), where—as Simon Schaffer comments—they "were to be ... what apples had allegedly been to gravitation".[52] Remarking that "if a Bubble be blown with Water first made tenacious by dissolving a little Soap in it, 'tis a common Observation, that after a while it will appear tinged with a great variety of Colours", Newton takes this phenomenon as a starting point both to disprove existing theories of colour and to develop his own theory that the full spectrum of colours is blended within and intrinsic to light itself.[53] Thus, rebutting

---

[50] Boyle, *Experiments and Considerations Touching Colours*, R1v-R2r (242–3).

[51] Hooke, *Micrographia*, I1v-2r (50–51). There were significant differences in Boyle's and Hooke's approaches to the phenomenon of colour, on which see Sabra, *Theories of Light*, 322. For cogent overview of Hooke's theories of light and colour, see Darrigol, *A History of Optics*, 53–57.

[52] Schaffer, "A Science Whose Business Is Bursting", 158. Elsewhere, Schaffer suggests that before their deployment by seventeenth-century natural philosophers, glass prisms—another object central to the *Opticks*—may have been used "as toys", noting that "there is some evidence that well into the seventeenth century prisms were seen as playfully deceitful and that the production of prismatic colours was indeed a common entertainment". Schaffer, "Glass Works", 73.

[53] Newton, *Opticks*, Book 2, part 1, Dd2v (2).

the Aristotelian theory that colours arise from the mixture of light and shadow, Newton offers a series of "Proof[s] by Experiments". Included amongst them, as "*Exper[iment]*. 4″, is the following passage:

> The Colours of Bubbles with which Children play are various, and change their Situation variously, without any respect to any Confine or Shadow. If such a Bubble be cover'd with a concave Glass, to keep it from being agitated by any Wind or Motion of the Air, the Colours will slowly and regularly change their situation, even whilst the Eye and the Bubble, and all Bodies which emit any Light, or cast any Shadow, remain unmoved. And therefore their Colours arise from some regular Cause which depends not on any Confine of Shadow.[54]

Here, an observation prompted by children playing is confirmed by further intervention, as a glass is used to temporarily arrest the bubble's ephemerality, verifying the broader principle that colours do not depend "on any Confine of Shadow". Rather, as Newton argues later in the *Opticks*, apparently white light is itself a compound of colours—a principle he proves by, amongst other things, an "Experiment" whereby water is "a little thicken'd with Soap" and then "agitated to raise a Froth". Subsequently, Newton observes, "after that Froth has stood a little, there will appear to one that shall view it intently various Colours every where in the Surfaces of the several Bubbles; but to one that shall go so far off, that he cannot distinguish the Colours from one another, the whole Froth will grow white with a perfect Whiteness".[55] Following a series of further observations and experiments on bubbles of various kinds, Newton concludes that colour is produced when "the transparent parts of Bodies, according to their several sizes, reflect Rays of one Colour, and transmit those of another, on the same grounds that thin Plates or Bubbles do reflect or transmit those Rays".[56]

   If Newton's famous but apocryphal assessment of his own career—"to myself I seem to have been only like a boy playing on the seashore, and diverting myself in now and then finding a smoother pebble or a prettier shell than ordinary, whilst the great ocean of truth lay all undiscovered before me"—describes youthful "diversions" as inconsequential distractions from the real business of contemplating "truth", then, his *Opticks*

---

[54] Newton, *Opticks*, Book 1, part 2, L3r (85).
[55] Newton, *Opticks*, Book 1, part 2, O4r (110).
[56] Newton, *Opticks*, Book 2, part 3, Hh4r (55).

Fig. 5.2 Pelagio Palagi, *Isaac Newton's Discovery of the Refraction of Light* (1827). Pinacoteca Tosio Martinengo, Brescia, Italy. © NPL—DeA Picture Library/Bridgeman Images

indicates a more generative role for play.[57] As with Bacon, Newton's observations of children playing with bubbles contribute to the formation and legitimation of a scientific theory, rather than retrospectively illustrating that theory once it is established. And for all their frothy fragility, bubbles—both those produced by the artless, aimless play of children, and those manufactured with intent by the natural philosopher—are sturdy enough to serve as the foundation or "grounds" of a whole new theory of light and colour.

Figure 5.2, a painting by the Bolognaise artist Filippo Pelagio Palagi depicting *Newton's Discovery of the Refraction of Light* (1827), crystallizes

---

[57] On these words, supposedly uttered not long before his death but found nowhere in his published works, see Gleick, *Isaac Newton*, 4.

Newton's insights—developed over the course of multiple experiments—as a single, momentary visual epiphany, even as it problematizes the significance of that epiphany in its evocation of the *vanitas* trope whereby bubbles represent the transience of human life and learning.[58] Palagi's painting both contrasts the severe, monumental, richly dressed figure of the philosopher with the joyful simplicity of the child and compares the two, both of whom sit (unlike the boy's nurse) in very similar poses, angled towards the right side of the painting with knees slightly apart and arms raised and bent at the elbows. The parallel is reinforced by the way both boy and man raise a hand and spread their fingers, the child reaching out to touch the physical bubble whilst Newton, as if startled by his own sudden apprehension, begins to grasp the idea the bubbles prompt.

*   *   *

Recording astronomical observations and constructing microscopes; harassing their older brothers for instruction in the latest discoveries; playing with pickled mackerel water; ingesting mandrake roots; chopping off lizards' tails; flaunting their flatulence; running and jumping and throwing; crafting pop guns; blowing soap bubbles: children at work and play flitter through the chronicles of seventeenth-century science. Sometimes treated as an object of scrutiny by older natural philosophers, in the "Ludicrous Experiments" that play generates we can also glimpse children acting, assiduously if not always intentionally, as dynamic experimental agents whose activities would provide pivotal to the subsequent history of science. Even critics of the Royal Society's approach found it hard to resist the allure of games: for all his suspicion of children's cognitive capacities, Descartes used examples drawn from their play as support for his physical theories. Outlining in a 1638 letter to Mersenne the contention "that bodies far from the centre of the earth do not weigh as much as those closer to it", he proffers as "evidence" not only "large birds such as cranes,

---

[58] On how "eighteenth-century popular depictions of Newton frequently reinforced the appealing idea that the philosopher developed his theory of light and colour through the childlike act of playing with soap bubbles", and on Palagi's painting as a "definitive" representation which blends "the neoclassical cult of Newton with the Romantic apotheosis of the child", see Kareem, "Enlightenment Bubbles", 89–90. See also Daston and Galison, *Objectivity*, 219.

LA PHYSIQUE.

Fig. 5.3 Bernard Picart, allegorical scene with *putti*, representing physics. Etching and engraving. From Bernard Le Bovier Fontenelle, *Oeuvres Diverses* (La Haye, 1729). © The Trustees of the British Museum

swans, etc.", which "fly much more easily when high in the air than when lower down", but also the example of "Paper kites flown by children".[59]

Between 1728 and 1729, there appeared from the presses of the Hague a three-volume work representing the *Oeuvres Diverses* [*Various Works*] of the philosopher and spokesman for the French Académie des Sciences, Bernard Le Bovier Fontenelle. Illustrated by the French Calvinist artist Bernard Picart, the *Oeuvres* include a series of images—two of which are included here, as Figs. 5.3 and 5.4—of chubby *putti*, engaged in a range of scientific pursuits, from natural history to geometry to anatomy to

[59] Descartes to Mersenne, 13 July 1638, in *The Philosophical Writings*, 3.112–13. John Ray refers to Descartes' observation in his *Three Physico-theological Discourses*, H4r-v (103–4): "One would think this were contrary to reason, that the lighter Air, such as is the superiour, should better support a weighty Body than the heavier, that is, the inferiour. Some imagine that this comes to pass by reason of the Wind which is constantly moving in the upper Air, which supports any Body that moves contrary to it. So we see that those Paper-kites which Boys make, are raised in the Air by running with them contrary to the Wind". Ray himself ultimately plumps for the alternative view "that the Gravity of Bodies decreases proportionably to their distance from the Earth".

L' ANATOMIE ET L' HISTOIRE NATURELLE.

Fig. 5.4   Bernard Picart, allegorical scene with *putti*, representing anatomy and natural history. Etching and engraving. From Bernard Le Bovier Fontenelle, *Oeuvres Diverses* (La Haye, 1729). © The Trustees of the British Museum

astronomy to physics.[60] In Fig. 5.3, for instance—representing physics— the *putti* cluster around a vacuum pump with a bird trapped inside, play at billiards, and dangle a set of magnetized keys. In Fig. 5.4, meanwhile— representing anatomy and natural history—they perform a dissection on a dog, observe a human foetus in a jar, and play with a spider. Arresting as they are, Picart's engravings offer a particularly charming example of a pervasive trend in sixteenth- and seventeenth-century scientific illustra- tion, where semi-celestial tots often frolic, collecting botanical specimens, peering through telescopes, performing chemical operations, and adding a note of whimsy to the otherwise austere frontispiece portraits of distin- guished figures such as Galileo Galilei and Sir Christopher Wren.[61]

---

[60] The playfulness of these images is especially intriguing given that Fontenelle himself "actively transformed lusus naturae into a category of ignorance rather than science", as Paula Findlen argues in "Ludic Postscript", 61; see also "Between Lent and Carnival", 266.

[61] See, for example, Francesco Villamena's portrait of Galileo, used as the basis for the frontispiece engravings in his "Istoria e dimostrazioni intorno alle macchie solari" (1613), *Saggiatore* (1623), and *Systema Cosmicum* (1635), in which one *putto* looks through a tele- scope, and another holds a quadrant and pen. Available at https://www.metmuseum.org/ art/collection/search/358902   and   https://cudl.lib.cam.ac.uk/view/PR-M-00010-

In many of these images, the *putti* serve several functions. Primarily, they have a classicizing purpose, as part a more general project to present experimentalism as part of a long and dignified elite tradition, rather than a form of base mechanical labour. Thus, commenting on an image from Caspar Schott's *Mechanica Hydraulico-Pneumatica* (1657), which shows *putti* operating Otto von Guericke's air pump, Steven Shapin argues that the "artistic convention" of depicting the "agents operating scientific instruments as putti, or cherubs, rather than human beings" was part of a broader tendency to occlude the roles played by lower-class technicians and servants who, in actuality, often carried out the physical work of experiment.[62]

In the frontispiece to a 1730 volume of two entomological works by the pioneering naturalist Maria Sybilla Merian—part of which is shown in Fig. 5.5—the *putti* have a slightly different role. Squabbling over sea-shells, inspecting botanical samples, and presenting Marian herself with baskets of fruit and flowers, these lovably chubby toddlers serve, as Alan Bewell comments, "to reduce the social anxieties raised by the unconventional aspects of Merian's life". Here, a woman whose ground-breaking entomological and botanical research took her as far as Dutch Suriname is placed firmly in domestic context where, in Bewell's words, she "looks more like a mother of sextuplets, than a natural history illustrator ... suggesting that ultimately her primary commitment as a female naturalist was to domestic family life, not the study of nature".[63] In fact, Merian—herself

---

00047-00001/1. See also Elisha Kirkall's portrait of Wren (c.1720–1742), available at https://www.npg.org.uk/collections/search/portrait/mw75211/Sir-Christopher-Wren. On "the ubiquity of putti in vignettes and frontispieces in all kinds of natural history [and] natural philosophy ... throughout the early modern period", see Hünniger, "Visible Labour?", 14. See also Heilbron, "Domesticating Science", on *putti* in Jesuit scientific illustrations, and Mûelenaere, "The Art of Learning", on the early origins of this convention. On *putti* as emblems of the importance (and ambivalence) of "learning through recreation", see Stafford, *Artful Science*, 42–4, 49.

[62] Shapin, "The Invisible Technician", 556. Similarly, Nick Wilding argues that during the seventeenth century, *putti* became "a visual euphemism for manual labour ... clearing the space of experimentation of labourers' bodies". Wilding, "Galileian Angels", 77. More recently, Hünniger has suggested that "a more complex history of putti at work emerges when taking the creators of the images as well as the collaborative nature of their productions more seriously. Manual as well as intellectual labour will appear more closely related" in "Visible Labour", 185.

[63] Bewell, "A Passion that Transforms", 38.

Fig. 5.5 Idealized portrait of Maria Sybilla Merian; detail from frontispiece to Merian, *De Europische Insecten ...*, bound with *Over de Voorttelingen Wonderbaerlyke Veranderingen der Surinaamsche Insecten* (Amsterdam, 1730). Etching and engraving, possibly hand-coloured by Merian and her daughters. Public Domain (Licensed as Attribution 4.0 International)

interested in insects from childhood onwards—relied heavily on the assistance of her daughters in her scientific work.[64] If, as this chapter has suggested, then, children were integral to the pursuit of early science, perhaps the distinction is less stark that Bewell implies. More generally, perhaps *putti* are present in these images not only as disguised and classicized manual technicians, but as playful representatives of the many and varied roles played by real children in the early modern natural and experimental sciences.

---

[64] Reitsma, *Maria Sibylla Merian and Daughters*. On how scientific "knowledge making" has sometimes been "a family business" throughout history, see Opitz *et al.*, "Introduction", 3.

CHAPTER 6

Conclusion: "This Philosophy of *Tops* and *Balls*"

Abstract This short Conclusion opens with a discussion of John Newbery's eighteenth-century scientific primer for children, *Newtonianism in the Nursery*. It goes on to explore the endurance of seventeenth-century ideas about childishness, experiment, and play into the eighteenth century and beyond, in the practice of modern science and in progressive educational theory. It speculates that in insisting on the isomorphism of experiment and play, members of the early Royal Society contributed to a reconfiguration in ideas about youth itself which still informs how we think about childhood and education today.

Keywords Newtonianism • Enlightenment • Disenchantment • Modern science • Experimental philosophy • John Newbery • Children • Play • Royal Society • Education • Albert Einstein • Blackawton Bees Project • Maria Edgeworth • Robert Boyle • Robert Hooke • Curiosity

In 1761, there appeared on the shelves of the enterprising London bookseller John Newbery, who specialized in works for children, a small gilt-covered volume with a grand title: *The Newtonian System of Philosophy*

107

E. L. Swann, *Science as Child's Play in Seventeenth-Century England*, https://doi.org/10.1007/978-3-031-75849-2_6

Adapted to the Capacities of Young Gentlemen and Ladies.[1] Providing a simplified summary of the new philosophy (for which, by this point, "Newtonianism" had become a byword) *The Newtonian System* is delivered in the form of lectures given by one "Tom Telescope"—a boy prodigy who combines scientific precocity with flawless moral conduct—to his young friends in the "Lilliputian Society".

Figure 6.1, from the 1798 edition, shows the diminutive Tom lecturing on motion, a topic he makes light work of with the help of a spinning top: "every boy who can whip his top knows what motion is as well as his master", he informs his audience.[2] As the lecture progresses, however, the limitations of naïve experience become clear when one of Tom's peers, the young Master Wilson, plays devil's advocate, objecting to Tom's avowal that "a body in motion will move for ever, unless some external cause stops it" with a sceptical question: "Shall any body tell me that my hoop or my top will run for ever, when I know by daily experience that they drop of themselves ...?" Master Wilson's reliance on "daily experience"—experience in the unfocused, quotidian, cumulative Aristotelian sense—is quickly dismissed by Tom, who nonetheless illustrates his counter-point not by turning away from "this Philosophy of *Tops* and *Balls*", but rather by scrutinizing play in more detail: "when you say nothing has touched the top or the hoop", he rebukes, "you forget their friction or rubbing against the ground the run upon, and the resistance they meet with from the air".[3]

A publishing phenomenon in its time, with nine editions in the eighteenth century and more produced well into the nineteenth century, *The Newtonian System*—which was probably authored by Newbery himself—kickstarted a trend for books which aimed to make the new philosophy of nature accessible to children.[4] Richly illustrated and engagingly

---

[1] Jim Secord comments that Newbery was "the first English publisher to make books specifically designed for children into a significant part of his business". Secord, "Newton in the Nursery", 129.

[2] Newbery, *The Newtonian System*, B3r (5). The work was based on John Locke's "Elements of Natural Philosophy", a manuscript treatise written by Locke in the late 1690s whilst acting as tutor to a twelve-year-old boy, and published posthumously; see Secord, "Newton in the Nursery", 131.

[3] Newbery, *The Newtonian System*, B3v-4r (6–7).

[4] Secord, "Newton in the Nursery", 139. On this work, and on "girls and boys" as "producers and consumers of a sensationalized knowledge" in the eighteenth century more generally, including a wide range of publications aimed at educating young people in science by entertaining them with "rational recreations", see Stafford, *Artful Science*, 50–71 (quotes at 51).

Fig. 6.1 Tom Telescope, from John Newbery, *The Newtonian System of Philosophy Adapted to the Capacities of Young Gentlemen and Ladies* (London, 1798). Copperplate. The Baldwin Library of Historical Children's Literatures, Special and Area Studies Collections, George A. Smathers Libraries, University of Florida. Public domain

conversational, its success was surely also driven, in part, by the appeal for children of Newbery's strategy of placing a young boy in a position of scientific authority. In the book, Tom's admiring audience includes fully-grown lords and ladies, as well as their children; *The Newtonian System* aimed to educate, but rather than talking down to its young readership, it presented them with an image of themselves as potential experts.

In offering to instruct "even the youngest children" in the principles of the "New Philosophy", James A. Secord suggests, *The Newtonian System* and its ilk represented a conspicuous departure from the past, for "in the seventeenth century, the Scientific Revolution was for adults only".[5] As this book has demonstrated, this was far from the case. Francis Bacon and his self-identified followers in the early Royal Society were ambivalent about childhood, sometimes seeing it, in line with Augustinian orthodoxy, as a condition of moral corruption and mental error, to be swiftly discarded in favour of manly self-mastery and mature judgement. At the same time, however, a forceful alternative stand of thought presented childhood, contrariwise, as an aspirational state to be cultivated. From this alternative perspective, children were crucial to the conceptualization and experience of the development of the new science, serving—amongst other things—as paradigms of sensory acuity and clarity; as instinctive natural historians and experimentalists; as unprejudiced possessors of an artless innocence which in some cases segued into something like the modern ideal of objectivity; and as representatives of a new form of experience prioritizing active and momentary intervention in the world over gradually accrued observations of it. Conversely, scientific experimentation and observation were themselves often considered as a form of play, with sophisticated sensory prosthetics such as the telescope figured as toys which opened the gateway to more youthful, pellucid form of perception. In this regard, experiment was predicated on childish innocence—but it also offered to restore it.

In fact, *The Newtonian System* itself draws examples from the previous century in its use of play to illustrate physical principles: in an echo of Bacon's *Novum Organum*, for instance, Tom demonstrates the "elastic principle in the air, which renders it capable of being rarified and condensed" using a "Pop-gun".[6] If anything, then, the winds of change were blowing in the opposite direction to that implied by Secord's assertion

---

[5] Secord, "Newton in the Nursery", 127.
[6] Newbery, *The Newtonian System*, E2r (39).

that before the eighteenth century, science was exclusively for grown-ups. As Second himself notes, "as the editions went into the early nineteenth century ... The audience no longer wanted to be lectured to by a child; therefore Tom grew up into a young man"; in the 1806 and 1812 editions young Master Tom disappears and is replaced by the fully grown "Mr Thomas Telescope", a lecturer for hire.[7] Is this, then, a story of disenchantment akin that propounded by Hans Blumenberg, and discussed in the introduction to this book: the gradual expulsion of the spirit of childish play from an increasingly austere, rational, adult science, simply deferred by a couple of centuries? "Enlightenment", runs Immanuel Kant's famous dictum, "is the human being's emergence from his self-incurred minority".[8] And it is assuredly the case that, today, many professional scientists would choose to stress the adult seriousness of their endeavours. As Susan A. Kirch and Michele Amoroso comment in their 2016 book *Being and Becoming Scientists Today*, "in our experience, using 'play' to describe science is typically not well received among our career-scientist colleagues who argue such a definition (one we find delightful) belittles the discipline and does not capture the nature and complexity of their work".[9]

In a spirit of playfulness, though, it is perhaps worth mentioning that what must be the most iconic modern image of a scientist—reproduced endlessly on posters, t-shirts, and mugs—shows the venerable Albert Einstein, in a photo taken by Arthur Sass on the occasion of the scientist's seventy-second birthday in 1951, exhibiting an expression of childlike glee, with his tongue protruding exuberantly from beneath his bushy white moustache.[10]

Indeed, discussions of the connections between science and play are rare but not unknown amongst modern scientists, with the chemist Pierre Laszlo calling on his peers in 2004 *American Scientist* article "not to be

---

[7] Secord, "Newton in the Nursery", 141. Tom was "re-established ... as a schoolboy" in the 1827 edition, but with his "age and maturity" emphasized.

[8] Kant, "What is Enlightenment?", 17.

[9] Kirch and Amoroso, *Being and Becoming Scientists*, 85.

[10] Dhaliwal, "Einstein's Tongue". Einstein's nostalgia for a period before a presumed expulsion of play from physics is evident in his foreword to a 1931 edition of the *Opticks*, where he (playfully) describes "Fortunate Newton" as the "happy childhood of science", and as the creator of "beautiful experiments which he ranged in order like playthings". Einstein, "Foreword", lix.

embarrassed to acknowledge that play is often what motivates us".[11] In the same piece, Laszlo recounts an anecdote shared by one "Nathan S. Lewis, a professor of chemistry at Caltech", who when working some years previously in the lab of a senior colleague, Harry Gray, found himself plagued by a coworker who "had the habit of going through their data and rushing to Gray with his interpretation". Lewis, quoted in Laszlo's article, recalls how he decided to teach the coworker a lesson:

> I manufactured an NMR [i.e. Nuclear Magnetic Resonance] spectrum that was a terrific result. We left it out as bait. [The coworker] took it and wrote up a paper on how important this result was. He was ready to go right to [the *Journal of the American Chemical Society*]. He had taken hook, line, and sinker on the manufactured piece of data. We didn't let him mail it, but we let him gloat around for a couple of days. This stopped him temporarily from taking our data and interpreting it before making sure it was right.[12]

Laszlo interprets this anecdote as evidence that "the playfulness of scientists", here expressed in the construction of a scientific hoax, can be "helpful … to the advancement of knowledge".[13] More specifically, what interests me here is the way that playfulness serves a temporal function of deferral or delay, reminding the coworker to avoid premature "interpretation". In encouraging the coworker to produce and assess his own data, rather than drawing unfounded conclusions based on others' work, Lewis' prank reminds him to linger with the evidence of his own senses, rather than leaping to theoretical speculation, in precisely the way that, many of the seventeenth-century authors discussed in the book suggest, children quite naturally do.

In the twentieth and twenty-first centuries, too, some so-called citizen science projects and science education initiatives have drawn on the contributions of children themselves, as producers—as well as recipients—of scientific knowledge.[14] Notably, in 2012, the neuroscientist Beau Lotto launched what he called the "Blackawton Bees Project". Working on the presumption that "science is play", Lotto's project enlisted twenty-five children from Blackawton Primary School in Devon, aged between eight and ten years old, in a three-month-long project on the visual behaviour of bumblebees, during which "they asked the questions, hypothesized the answers, designed the games (in other words, the experiments) to test these hypotheses and analysed the data".[15] The resulting paper, published in the peer-reviewed Royal Society journal *Biology Letters*, was formulated by the children but "transcribed" by Lotto. That the children took "their own observations of the world" rather than "the scientific literature" as the basis for their research, Lotto claims, "does not diminish the resulting data, scientific methodology, or merit of the discovery ... on the contrary, it reveals science in its truest (most naive) form".[16]

For all its shiny professionalism, seriousness, and technological heft, then, modern science has not quite expelled childish play entirely. The experimental child, moreover, lives on in the realm of pedagogy: today, the "active" or "child-centred" forms of learning that hold sway in much of Europe and the United States bear the imprint of Enlightenment educational theory, itself shaped by engagement with earlier ideas—in some cases lifted directly from the work of the natural philosophers discussed here—about the scientific inclinations and capacities of children. Tracing all the ways in which this is the case would take us beyond the scope of the present book, and is a topic ripe for further research. For one example, however, we may turn to the Anglo-Irish daughter-and-father duo Maria and Richard Lovell Edgeworth's *Practical Education* (1798), a work which had a significant impact on later advocates of progressive and play-centred education, including the German pedagogue Friedrich Froebel, who established the first "kindergarten" in 1837, and the twentieth-century Italian physician Maria Montessori, whose famous "method" emphasizes the cultivation of children's natural curiosity through direct, hands-on learning.[17]

---

[15] Lotto, "Why Science is Like Play", 11 November 2012, https://edition.cnn.com/2012/11/11/opinion/lotto-ted-science-play/index.html; Lotto, "Background", in Blackawton *et al.*, "Blackawton Bees", 168.

[16] Lotto, "Background", in Blackawton et al., "Blackawton Bees", 168.

[17] *Practical Education* is now "thought to be primarily Maria's work"; see Rendall, "'Elementary Principles'", 620. For the "Montessori Method", see Montessori, *The Montessori Method*, especially "Nature in Education". On how "the ideas of ... the Edgeworths and Froebel still resonate widely and underpin the progressive education movement that has influenced our modern education system in so many ways", see Richardson, "How We Learned to Teach"; see also Doddington and Hilton, *Child Centred Education*, 6. On how *Practical Education* "anticipated" the ideas of Froebel and Montessori, see Lyons, "Play and Toys", 312, 315.

The influence of Locke and Rousseau on the Edgeworths' ideas about education is often noted.[18] Less frequently acknowledged is their engagement with the earlier generation of experimental philosophers who stood behind those figures. Rejecting rote learning and strict disciplinarianism, the Edgeworths—themselves keen amateur scientists—advise that "practical education begins very early, even in the nursery", describing children as instinctive natural historians and experimentalists in a way redolent of (and sometimes explicitly citing) earlier Royal Society rhetoric. "When children are busily trying experiments upon objects within their reach", they warn, "we should not, by way of saving them trouble, break the course of their ideas, and totally prevent them from acquiring knowledge by their own experience": the infant engaged in "trying the difference between pushing and pulling, rolling or sliding, the powers of the wedge or the lever" is engaged not in idle play but a process of learning.[19] Further illustration of this principle is given in the form of an anecdote about "a little boy of nine years old", identified only by his initial, "S.——", who:

> was standing without any book in his hand, and seemingly idle; he was amusing himself with looking at what he called a rainbow upon the floor; he begged his sister M—— to look at it; then he said he wondered what could make it; how it came there. The sun shone bright through the window; the boy moved several things in the room, so as to place them sometimes between the light and the colours which he saw upon the floor, and sometimes in a corner of the room where the sun did not shine. As he moved the things, he said, 'This is not it;' 'nor this;' 'this has'n't any thing to do with it'. At last he found, that when he moved a tumbler of water out of the place where it stood, his rainbow vanished. ... He emptied the glass; the colours remained, but they were fainter. S—— immediately observed, that it was the water and glass together that made the rainbow. "But", said he, 'there is no glass in the sky, yet there is a rainbow, so that I think the water alone would do, if we could but hold it together without the glass. Oh I know how I can manage.' He poured the water slowly out of the tumbler into a basin, which he placed where the sun shone, and he saw the colours on the floor twinkling behind the water as it fell: this delighted him much; but he asked why it would not do when the sun did not shine.

[18] See, for instance, Whitbread, *The Evolution of the Nursery Infant-School*, 18; Myers, "'Anecdotes from the Nursery'", 228–8; Lyons, "Play and Toys", 312; and Doddington and Hilton, *Child Centred Education*, 6; 228–9.

[19] Edgeworth and Edgeworth, *Practical Education*, vol. 1, C1r (9).

A rigid preceptor, who thinks that every boy must be idle who has not a Latin book constantly in his hand, would perhaps have reprimanded S—— for wasting his time *at play*, and would have summoned him from his rainbow to his *task*; but it is very obvious to any person free from prejudices, that this child was not idle whilst he was meditating upon the rainbow on the floor; his attention was fixed; he was reasoning; he was trying experiments.[20]

In this passage, the Edgeworths describe the child's patient, pleasurable investigations into the refraction of light in a way with recalls not only Newton with his soap bubbles and prisms but also the work of Descartes. "We may call this *play* if we please", the Edgeworths assert crisply, "and we may say that Descartes was at play, when he first verified Antonio de Dominis bishop of Spalatros treatise of the rainbow, by an experiment with a glass Globe". Carefully distinguishing between "the powers of reasoning, or the abilities of the child and the philosopher", they nonetheless claim that "the same species of attention was exerted by both".[21] Elsewhere in *Practical Education* they suggest that toy shops should sell:

cheap microscopes, which will unfold a world of new delights to children … they may easily be led to try slight experiments in optics, which will, at least, give the habits of observation and attention. … We may point out that great discoveries have often been made by attention to slight circumstances. The blowing of soap bubbles, as it was first performed as a scientific experiment by the celebrated Dr. Hook, before the Royal Society … may be read to children, and they will be pleased when they observe that what at first appeared only a trifling amusement, has occupied the understanding, and excited the admiration, of some great philosophers.[22]

Here, the microscope-as-toy is designed not to enable children to make new discoveries, but rather to inculcate "habits of observation and attention". Nonetheless, the reader, as much as the notional children the text describes, is encouraged to consider both the ways in which play functions

---

[20] Edgeworth and Edgeworth, *Practical Education*, vol. 1, H4r-I1r (55–57).

[21] Edgeworth and Edgeworth, *Practical Education*, vol. 1, I1r (57). The reference to Descartes' verification of Antonio de Dominis' theory of the rainbow comes from Priestley, *The History*, 1.51.

[22] Edgeworth and Edgeworth, *Practical Education*, vol. 1, E3v-4r (30–31).

as a form of experimental learning and the ways in which experiment itself, even when practiced by such a towering figure as "the celebrated Dr. Hook", may have its roots in the kinds of "trifling amusement[s]" that occupy the leisure time of children.

The interest of Hooke and other members of the Royal Society—as well as its critics—in the childish impulse to accumulate natural ephemera and oddities, too, resonates in *Practical Education*. "Natural history", the Edgeworths assert, "interests children at an early age", and this disposition should be fostered with the provision of "empty shelves in their cabinets, to be filled with their own collections".[23] Whereas, however, earlier discussions of children as instinctive collectors either derided or lauded their disregard of worldly standards of value in choosing their treasures, the Edgeworths emphasize instead the inculcation of a taxonomical impulse. Here, the child's "enthusiasm" for collecting "a variety of pebbles and common stones, which they value as great curiosities", is laudable not because it has something to teach the adult naturalist about the beauty and worth of apparently insignificant natural objects, but rather because it serves the educational purpose of teaching children themselves "how to direct their researches", enabling them to "acquire a taste for order by the best means".[24]

"Based originally on Lockean ideas", write Christine Doddington and Mary Hilton in their history of modern child-centred education, "*Practical Education* was the first educational work to fully configure an experimental and holistic method of 'discovery' in education ... in its pages we can trace a direct route back from our child-centred ideas to their Enlightenment origins".[25] As I have suggested, however, in laying the groundwork for modern, progressive educational theory, the Edgeworths did not just deploy the work of Locke; rather, they drew on a vibrant vein of thinking in the work of the early Royal Society regarding the richly isomorphic relationship between play and experiment. If the seventeenth-century conceptualization of experiment as child's play is treated with some suspicion by many professional scientists in the twenty-first century, then, the inverse idea—that child's play is a form of experimental learning—lives on, albeit in a somewhat attenuated form, in educational theory and practice. And to the extent that members of the early Royal Society helped—via the

---

[23] Edgeworth and Edgeworth, *Practical Education*, vol. 1, E2v-E3r (28–9).
[24] Edgeworth and Edgeworth, *Practical Education*, vol. 1, E3r-v (29–30).
[25] Doddington and Hilton, *Child Centred Education*, 6.

work of the Edgeworths and others—to shape that theory and practice, they also contributed to a transformation in how we think about childhood and education today.

* * *

In *Iconoclasm as Child's Play*, Joe Moshenska recommends that the early modern intertwinement of play with natural philosophy should be "borne in mind whenever we encounter one of the large-scale narratives that see play as depleted or disappearing under modernity", blurring "a clear distinction between grim technological domination and winsome, playful forms of practice".[26] In one way or another, as Moshenska also points out, the history and philosophy of early experimentalism has often accentuated entanglements of scientific knowledge and nefarious power. This tendency informs, amongst other things, Horkheimer and Adorno's indictment of Baconian instrumental reason as a contributing factor in the burgeoning of twentieth-century fascism, as "what human beings seek to learn from nature is how to use it to dominate wholly both it and human beings", and the work of ecocritical and feminist historians including Carolyn Merchant, for whom Bacon's rhetorical depictions of experiment as a masculinized and aggressive domination of a personified, feminized nature reflect science's sexist, objectifying, and violent practices.[27] Often quoting Bacon's own assertion, *scientia potestas est* ("knowledge is power"), or his declaration that "nature must be taken by the forelock", such accounts vary in emphasis, but concur in framing early science as complicit in broader structures of oppression and exclusion.[28]

Children, of course, are not exempt from the impulse to dominate; as the nonconformist minister and educator Hezekiah Woodward writes in his 1641 *A Light to Grammar*, "Children delight in the pain and vexation

---

[26] Moshenska, *Iconoclasm as Child's Play*, 181–2.

[27] Horkheimer and Adorno, *Dialectic of Enlightenment*, 2; Merchant, *The Death of Nature*. The idea that Bacon thought of experiment as a form of torture has been challenged by historians including Peter Pesic, who suggests that he considered it not as a unilateral exercise of power, but as a mutual struggle whereby the experimenter was tested as much as the matter he worked on. Pesic, "Wrestling with Proteus".

[28] Bacon, *Religious Meditations* [*Meditationes Sacrae*], E3v. Bacon refers here to God's knowledge and power, but in other works he extends the principle to the realm of human knowledge: "Human knowledge and power come to the same thing", he writes in the *Novum Organum*, 11.65.

of those weake creatures, that are in their power".[29] Indeed, according to
the Augustinian model of childhood as a state of unregenerate sin, they
may be more, not less, inclined than adults to pursue their own wilful and
sometimes cruel desires. And if the experimental philosophers in this book
often chose to emphasize childish innocence, they also acknowledged this
tendency. "The Study of Physiologie", writes Boyle in *Some Considerations
Touching the Usefulnesse of Experimental Naturall Philosophy*,

> is not only Delightful, as it teaches us to Know Nature, but also as it teaches
> us in many Cases to Master and Command her. ... And how Naturally we
> affect the Exercise of this Power over the Creatures may appear in the
> Delight Children take to do many things ... that seem to proceed from an
> Innate Propensity to please themselves in imitating or changing the
> Productions of Nature.[30]

For Boyle, desire for power "over the Creatures" via the acquisition of
natural knowledge is legitimized by reference to childish play, which similarly stems from an experimental impulse to intervene in, alter, and master
"the Productions of Nature". Hooke marvelled at the "surpassing beauty"
of the flies' wings viewed under the microscope, "variegated with all the
variety of curious bright and vivid colours imaginable".[31] But he also cut
their heads off; a move reminiscent of the "wanton boys" described by
Gloucester in Shakespeare's *King Lear*, who (just as the Gods play merci-
lessly with human lives) kill flies for "sport", or Valeria's description of
how young Martius "mammocked" a butterfly in his *Coriolanus*
(c.1608–1609; 1623).[32] Play, then, is not (or not only) a "winsome",
ameliorative counterbalance to "grim technological domination", but can
also itself take the form of a naturalized and naturalizing manifestation of
the will to power. Despite this qualification, however, the narratives of
domination that have themselves dominated the historiography of early
modern science are surely due a rethink, for—as I have suggested throughout this book—any authority experimental knowledge offers is grounded

[29] Woodward, *A Childes Patrimony*, H2v. Katherine Larson discusses "play's potential for
cruelty and violence" in "'Certein childeplayes'", 77–78.

[30] Boyle, *Some Considerations Touching the Usefulness*, D2r (19).

[31] Hooke, *Micrographia*, Aa3v (174).

[32] Hooke, *Micrographia*, Aa4r (175); Shakespeare, *King Lear*, 4.1.37–38; Shakespeare,
Coriolanus, 1.3.61.

in a profound sensory receptivity which may recall the vulnerability, as well as the innocence, of youth.

Let us end, then, not with the child's casual cruelty but with their curiosity. As David Carroll Simon has shown, for all the new scientist's interest in power and domination, "equally important to this moment in intellectual history is the beguilingly simple intuition that the world is different from what anyone expects".[33] Tracing the importance of what he calls "the observational mood"—an experience of alert but relaxed, digressive, and pleasurable openness to the world—to the development of early science, Simon contends that "England's scientific revolution gathers momentum … from experiences of languorous drift".[34] This mood of "drift"—and the revelatory encounters with the world's unexpectedness that it sometimes enables—is, I would add, also experienced by men like Bacon and Boyle as a return to the passivity of childhood: a stage of life when a state of unfazed bewilderment may be habitual; when the pleasurable byways of play beckon alluringly but without the urgency of adult commitments; and wonder is no sudden eruption but a constant, unexceptional, radiant companion. The child, writes Hezekiah Woodward, is:

> here and there, and every where with his sticke, or with his gun, or with his casting stones; perhaps if these be not at hand, he is blowing up a feather. … This child is desirous after knowledge, very curious and enquiring that way.[35]

Recalling, in his spirited but undirected roaming "here and there", the "boy that run up and down the room" in Samuel Pepys' description of Margaret Cavendish's visit to Arundell House, the child's thirst for knowledge expresses itself not in focused investigation of a single phenomenon but in a haphazard, lackadaisical form of play that employs whatever ephemera he finds "at hand". It may seem a long way from this lone, unruly figure to the monumental triumphs and catastrophes of modern science: from sticks and stones to quantum computing and the large hadron collider, from a child blowing a feather in the air to wind turbines. Perhaps, though, the distance is not quite as far as we have previously thought.

---

[33] Simon, *Light without Heat*, 21.
[34] Simon, *Light without Heat*, 18.
[35] Woodward, *A Light to Grammar*, C5v-6r (26–7).

Bibliography

Primary

Agrippa, Henry Cornelius. *Three Books of Occult Philosophy.* Translated by J.F. London, 1651.

Anon. "An Extract of a Letter Concerning an Optical Experiment, Conducive to a Decayed Sight". *Philosophical Transactions* 3.37 (1668): 727–731.

Aquinas, Thomas. *Summa Theologiæ.* Translated by Fathers of the English Dominican Province. Revised Edition, 1920. https://www.newadvent.org/summa/3168.htm, edited by Kevin Knight. Accessed 01 November 2022.

Aristotle. *Metaphysics.* Edited and translated by W. D. Ross. Oxford University Press, 1953.

Aristotle, *Nicomachean Ethics.* Edited by Lesley Brown. Translated by David Ross. Oxford University Press, 2009.

Ascham, Roger. *The Scholemaster.* London, 1570.

Aristotle, *De anima* [*On the Soul*]. Edited and translated by Christopher Shields. Oxford University Press, 2016.

Augustine. *Confessions.* Translated by William Watts. London, 1631.

Augustine, *A Treatise on the Origin of the Human Soul, Addressed to Jerome.* Translated by J.G. Cunningham. In *Nicene and Post-Nicene Fathers*, vol. 1. Edited by Philip Schaff. Christian Literature Publishing Co., 1887. Revised and edited for New Advent by Kevin Knight at http://www.newadvent.org/fathers/1102166.htm. Accessed 18 June 2022.

Bacon, Francis. *Of the Advancement and Proficience of Learning* [*De Augmentis Scientiarum*]. Oxford, 1640.

Bacon, Francis. *The Advancement of Learning*. In *The Oxford Francis Bacon*, vol. 4. Edited by Michael Kiernan. Oxford University Press, 2000a.

Bacon, Francis. *The Essayes or Counsels, Civill and Morall*. 2nd ed. London, 1625.

Bacon, Francis. *Instauratio Magna*. In *The Oxford Francis Bacon*, vol. 11. Edited by Graham Rees and Maria Wakely. Oxford University Press, 2004a.

Bacon, Francis. *Natural and Experimental History*. In *The Oxford Francis Bacon*, vol. 12. Edited by Graham Rees and Maria Wakely. Oxford University Press, 2007.

Bacon, Francis. *A New Abecedarium of Nature [Abecedarium Novum Naturae]*. In *The Oxford Francis Bacon*, vol. 13. Edited by Graham Rees. Oxford University Press, 2000b.

Bacon, Francis. *Novum Organum*. In *The Oxford Francis Bacon*, vol. 11. Edited by Graham Rees and Maria Wakely. Oxford University Press, 2004b.

Bacon, Francis *"Phenomena of the Universe"* [*Phaenomena Universi*]. In *The Oxford Francis Bacon*, vol. 6, edited by Graham Rees. Oxford University Press, 1996.

Bacon, Francis. "Promus of Formularies and Elegancies". In *The Oxford Francis Bacon*, vol. 1. Edited by Alan Stewart with Harriet Knight. Oxford University Press, 2012.

Bacon, Francis. *Preparative to a Natural History*. In *The Oxford Francis Bacon*, vol. 11. Edited by Graham Rees and Maria Wakely. Oxford University Press, 2004c.

Bacon Francis. *Religious Meditations* [*Meditationes Sacrae*]. In *Essayes*. London, 1597.

Bacon, Francis. *Sylva Sylvarum, or, A Natural History in Ten Centuries*. London, 1658.

Bacon, Francis. *The Works of Francis Bacon*, vol. 7. Edited by James Spedding, Robert Leslie Ellis and Douglas Denon Heath. London, 1859.

Bartholomaeus, Angelicus. *Batman uppon Bartholome his Booke De proprietatibus rerum*. Translated by Stephen Batman. London, 1538.

Baxter, Richard. *A Christian Directory*. London, 1673.

Beale, John. "An Experiment to Examine, what Figure, and Celerity of Motion Begetteth, or Encreaseth Light and Flame". *Philosophical Transactions* 1.13 (1666): 1226-228.

Beale, John, "Two Instances of Something Remarkable in Shining Flesh". *Philosophical Transactions* 11.125 (1676): 599-603.

Boyle, Robert. "An Account of Philaretus in his Minority". In *Robert Boyle by Himself and His Friends*, edited by Michael Hunter. Routledge, 1994.

Boyle, Robert. *Certain Physiological Essays*. London, 1661.

Boyle, Robert. *The Excellency of Theology Compar'd with Natural Philosophy*. London, 1674.

Boyle, Robert. *Experiments and Considerations Touching Colours*. London, 1664.

Boyle, Robert. *New Experiments and Observations Touching Cold*. London, 1665a.

Boyle, Robert. *Occasional Reflections upon Several Subjects*. London, 1665b.

Boyle, Robert. *Some Considerations about the Reconcileableness of Reason and Religion*. London, 1675.

Boyle, Robert. *Some Considerations Touching the Usefulnesse of Experimental Naturall Philosophy*. Oxford, 1663.

Brahe, Tycho. *Instruments of the Renewed Astronomy*. Edited and translated by Alena Hadravová, Petr Hadrava, and Jole R. Shackelford. Koniasch Latin Press, 1996. [Revised translation based on *Tycho Brahe's Description of his Scientific Instruments and Work*. Edited and translated by Hans Raeder, Elis Strömgren, and Bendt Strömgren. Ejnar Munksgaard, 1946.]

Brathwaite, Richard. *The English Gentleman*. London, 1631.

Bullivant, Benjamin. "Part of a Letter from Mr. Benjamin Bullivant, at Boston, in New England; to Mr. James Petiver, Apothecary, and Fellow of the Royal Society, in London". *Philosophical Transactions of the Royal Society of London* 20.240 (1698): 167-68.

Burnet, Gilbert. "The Burnet Memorandum". In *Robert Boyle by Himself and His Friends*, edited by Michael Hunter. Routledge, 1994.

Burnet, Gilbert. *A Sermon Preached at the Funeral of the Honourable Robert Boyle*. London, 1692.

Butler, Samuel. "The Elephant in the Moon [in Short Verse]". In *The Genuine Remains in Verse and Prose of Mr. Samuel Butler*, vol. 1. Edited by R. Thyer. London, 1759.

Calvin, John. *A Harmonie Upon the Three Evangelists, Matthew, Mark and Luke*. London, 1584.

Calvin, John. *The Institution of Christian Religion*. Translated by Thomas Norton. London, 1561.

Cavendish, Margaret. *Ground of Natural Philosophy*. London, 1668.

Cavendish, Margaret. *Observations Upon Experimental Philosophy*. London, 1666.

Centlivre, Susannah. *The Basset-table: A Comedy*. London, 1706.

Charleton Walter. *Physiologia Epicuro-Gassendo-Charltoniana, or, A fabrick of Science Natural*. London, 1654.

Childrey, Joshua. *Britannia Baconica: or, The Natural Rarities of England, Scotland, & Wales*. London, 1662.

Comenius, Jan Amos [Johann Amos Komenský]. *Orbis Sensualium Pictus*. London, 1659.

Comenius, Jan Amos [Johann Amos Komenský], and Peter Colbovius. *Sendschreiben des Petrus Colbovius an J. A. Comenius (1650) und Brief des J. A. Comenius an Colbovius (1650): eine pädagogische Korrespondenz aus dem 17.Jahrhundert*. Aloys Henn Verlag, 1974.

Cowley, Abraham. *A Proposition for the Advancement of Experimental Philosophy*. London, 1661.

Cowley, Abraham. "To the Royal Society". In Thomas Sprat, *The History of the Royal Society of London*. London, 1667.

Cuff, Henry. *The Differences of the Ages of Man's Life*. London, 1607.

Cummings, Brian, ed. *The Book of Common Prayer: The Texts of 1549, 1559, and 1662*. Oxford University Press, 2011.

Dante, Alighieri. *The Divine Comedy: Paradise,*. Edited by D. L. Sayers. Translated by B. Reynolds. Penguin, 1962.

Dee, John. *A True & Faithful Relation of What Passed for Many Yeers Between Dr. John Dee... and Some Spirits*. London, 1659.

Denison, John. *The Christians Care for the Soules Safety*. London, 1621.

Descartes, René. *Meditations on First Philosophy: With Selections from the Objections and Replies*. Edited and translated by John Cottingham. 2nd ed. Cambridge University Press, 2017a.

Descartes, René. *The Philosophical Writings of Descartes*, vol. 1. Translated by John Cottingham, Robert Stoothoff, and Dugald Murdoch. Cambridge University Press, 1985.

Descartes, René. *The Philosophical Writings of Descartes*, vol. 3, *The Correspondence*. Translated by John Cottingham, Robert Stoothoff, Dugald Murdoch, and Anthony Kenny. Cambridge University Press, 1991.

Descartes, René. *Principles of Philosophy*. www.earlymoderntexts.com. Version by Jonathan Bennett, 2017b.

Drake, Judith [attributed]. *An Essay in Defence of the Female Sex*. London, 1696.

Dury, John. *The Reformed School*. London, 1649.

Earle, John. *Micro-cosmographie, or A Piece of the World Characteriz'd*. London, 1628.

Edgeworth, Maria, and Richard Lovell Edgeworth. *Practical Education*, vol 1. London, 1798.

Erasmus, Desiderius. *That Chyldren Oughte to be Taught and Brought up Gētly in Vertue and Learnynge* [*The Education of Children*]. In (and translated by) Richard Sherry, *A Treatise of Schemes and Tropes*. London, 1550.

Erasmus, Desiderius. *The Civilite of Childehode*. Translated by Thomas Paynell. London, 1560.

Fontenelle, Bernard Le Bovierde. *Oeuvres Diverses* [*Various Works*]. Gosse & Neaulme, 1728-1729.

Glanvill, Joseph. *The Vanity of Dogmatizing*. London, 1661.

Goodwin, Thomas. *The Vanity of Thoughts Discovered*. London, 1638.

Hooke, Robert. "An Account of the Plant, call'd Bangue". In *Philosophical Experiments and Observations of the Late Eminent Robert Hooke*, edited by W. Derham. London, 1726a.

Hooke, Robert. *The Diary of Robert Hooke, 1672–1680*, edited by Henry W. Robinson and Walter Adams. Taylor & Francis, 1935.

Hooke, Robert. "Discourse Concerning Telescopes and Microscopes; With a Short Account of their Inventors". In *Philosophical Experiments and Observations of the Late Eminent Robert Hooke*, edited by W. Derham. London, 1726b.

Hooke, Robert. "A General Scheme, or Idea of the Present State of Natural Philosophy and How its Defects May be Remedied", in *The Posthumous Works of Robert Hooke*, edited by Richard Waller. London, 1705.

Hooke, Robert. *Micrographia, or Some Physiological Descriptions of Minute Bodies Made by Magnifying Glasses*. London, 1665.

Jonson, Ben. *Discoveries*, edited by Lorna Hutson. In *The Cambridge Edition of the Works of Ben Jonson*, vol. 7. Edited by David Bevington, Martin Butler, and Ian Donaldson. Cambridge: Cambridge University Press, 2012.

Kant, Immanuel. "An Answer to the Question: What is Enlightenment?" In *Practical Philosophy*. Edited and translated by Mary J. Gregor. Cambridge University Press, 1996.

Kempe, William. *The Education of Children in Learning*. London, 1588.

Kennett, White. *Parochial Antiquities*. Oxford, 1695.

King, William. *The Transactioneer, With Some of his Philosophical Fancies in Two Dialogues*. London, 1700.

L'Estrange, Roger. *Senecas Morals Abstracted*. London, 1679.

Linnaeus, Carl. *Diary*. Translated by C. Troilius. In Richard Pulteney, *A General View of the Writings of Linnaeus*, edited by William George Maton. 2nd ed. J. Mawman, 1805.

Locke, John. *An Essay Concerning Human Understanding*, edited by P. H. Nidditch. Oxford University Press, 1975.

Locke, John. *Some Thoughts Concerning Education*. In *The Educational Writings of John Locke*, edited by John William Adamson. Cambridge University Press, 2011.

Malebranche, Nicolas. *Malbranche's Search after Truth*. Translated by Thomas Taylor. London, 1694.

Martin, Benjamin. *The Young Gentlemen's and Ladies Philosophy*. London, 1759.

Montaigne, Michel de. *Essays*. Translated by John Florio. London, 1613.

Montessori, Maria. *The Montessori Method: Scientific Pedagogy as Applied to Child Education*. Translated by Anne E. George. Barnes and Noble, 2003 [1912].

Murphy, Kathryn, and Anita Traninger, eds., *The Emergence of Impartiality*. Brill, 2014.

Nedham, Marchamont. *Medela Medicinæ*. London, 1665.

Newbery, John [?]. *The Newtonian System of Philosophy Adapted to the Capacities of Young Gentlemen and Ladies*. London, 1761.

Newton, Isaac. *Opticks, or a Treatise of the Reflexions, Refractions, Inflexions, and Colours of Light*. London, 1704.

Pepys, Samuel. *The Diary of Samuel Pepys*, vol 5. Edited by Robert Latham and William Matthews. Harper Collins, 2000a.

Pepys, Samuel. *The Diary of Samuel Pepys*, vol 8. Edited by Robert Latham and William Matthews. Harper Collins, 2000b.

Pepys, Samuel. *The Diary of Samuel Pepys*, vol 9. Edited by Robert Latham and William Matthews. Harper Collins, 2000c.

Pliny the Elder. *Natural History*, vol. 1. Translated by H. Rackham. Loeb Classical Library 330. Harvard University Press, 1938.

Plato. *Meno*. In *Laches. Protagoras. Meno. Euthydemus*. Translated by W. R. M. Lamb. Loeb Classical Library 165. Harvard University Press, 1924.

Plato. *Timaeus*. In *The Dialogues of Plato*, vol. 3. Translated by Benjamin Jowett. Oxford University Press, 1892.

Priestley, Joseph. *The History and Present State of Discoveries Relating to Vision, Light, and Colours*. London, 1772.

Ray, John. *Further Correspondence of John Ray*, edited by Robert W. T. Gunther. London, 1928.

Ray, John. *Observations Topographical, Moral, & Physiological; Made in a Journey through Part of the Low-countries, Germany, Italy, and France*. London, 1673.

Ray, John. *Three Physico-theological Discourses*. London, 1693.

Ray, John. *The Wisdom of God Manifested in the Works of the Creation*. London, 1691.

Reid, Thomas. *An Inquiry into the Human Mind on the Principles of Common Sense*. Cambridge University Press, 2011.

Ringler, William. "*Poeta Nascitur Non Fit*: Some Notes on the History of an Aphorism." *Journal of the History of Ideas* 2. 4 (1941): 497–504.

Rousseau, Jean-Jacques. *Emile*. Translated by Allan Bloom. Basic Books, 1979.

Rzepka, Adam. "'Rich eyes and poor hands': Theaters of Early Modern Experience". In *Knowing Shakespeare: Senses, Embodiment and Cognition*, edited by Lowell Gallagher and Shankar Raman. Palgrave Macmillan, 2010.

Scheiner, Christophe. *Oculus, sive Fundamentum Opticum*. London, 1652.

Shadwell, Thomas. *The Virtuoso*. London, 1676.

Shakespeare, William. *Coriolanus*, edited by Peter Holland. The Arden Shakespeare. Bloomsbury Publishing, 2013.

Shakespeare, William. *King Henry VI, Part 1*, edited by Edward Burns. The Arden Shakespeare. Thomson Learning, 2000.

Shakespeare, William. *King Lear*, edited by R. A. Foakes. The Arden Shakespeare. Thomson Learning, 1997.

Sheffield, John. *The Rising Sun, or, The Sun of Righteousnesse*. London, 1654.

Sherburne, Edward. *The Sphere of Marcus Manilius Made an English Poem*. London, 1675.

Simpson, William. *Philosophical Dialogues Concerning the Principles of Natural Bodies*. London, 1677.

Sprat, Thomas. *The History of the Royal Society of London*. London, 1667.

Taylor, Jeremy. *Unum Necessarium. Or, the Doctrine and Practice of Repentance*. London, 1655.

Traherne, Thomas. *Centuries of Meditations*. In *The Works of Thomas Traherne*, vol. 5. Edited by Jan Ross. Boydell and Brewer, 2013.

Traherne, Thomas. *Commentaries of Heaven*. In *The Works of Thomas Traherne*, vol. 4. Edited by Jan Ross. Boydell and Brewer, 2007.

Traherne, Thomas. "The Preparative". In *The Works of Thomas Traherne*, vol. 6. Edited by Jan Ross. Boydell and Brewer, 2014.

Underhill, Henry. "An Account of What Happened to a Child on Swallowing Two Copper Farthings". *Philosophical Transactions* 246 (1698): 424.

Waller, Richard. "The Life of Dr. Robert Hooke", in Robert Hooke, *The Posthumous Works of Robert Hooke*, edited by Richard Waller. London, 1705.

Ward, John. *The Lives of the Professors of Gresham College*. London, 1740.

Webster, John. *Academiarum Examen, or, The Examination of Academies*. London, 1654.

Willughby, Francis. *Francis Willughby's Book of Games: A Seventeenth-Century Treatise on Sports, Games and Pastimes*, edited by David Cram, Jeffrey L. Forgeng, and Dorothy Johnston. Routledge, 2016.

Willughby, Francis. *The Ornithology*. Edited and translated by John Ray. London, 1678.

Woodward, Hezekiah. *A Light to Grammar, and All Other Arts and Sciences*. London, 1641.

Woodward, Hezekiah. *A Childes Patrimony*. London, 1640.

Wright, Thomas. *The Passions of the Mind*. Enlarged ed. London, 1604.

SECONDARY

Aït-Touati, Frèdèrique. *Fictions of the Cosmos: Science and Literature in the Seventeenth Century*. Translated by Susan Emanuel. University of Chicago Press, 2011.

Allen, Barry. *Empiricisms: Experience and Experiment from Antiquity to the Anthropocene*. Oxford University Press, 2020.

Allen, Don C. *Doubt's Boundless Sea: Skepticism and Faith in the Renaissance*. John Hopkins University Press, 1964.

Almond, Philip C. "The Journey of the Soul in Seventeenth-Century English Platonism". *History of European Ideas* 13.6 (1991): 775-91.

Ariès, Phillipe. *Centuries of Childhood: A Social History of Family Life*. Translated by Robert Baldick. Alfred A. Knopf, 1962.

Aulakh, Pavneet. "'Small Things Discover Great': 'Lower Wisdom' in *Paradise Lost*". In *Milton and the New Scientific Age: Poetry, Science, Fiction*, edited by Catherine Gimelli Martin. Routledge, 2019.

Balakier, James J. "Thomas Traherne's Dobell Series and the Baconian Model of Experience". *English Studies* 70.3 (1989): 233-247.

Benedict, Barbara M. *Curiosity: A Cultural History of Early Modern Inquiry*. University of Chicago Press, 2002.

Bewell, Alan. "A Passion that Transforms: Picturing the Early Natural History Collector". In *Figuring it Out: Science, Gender, and Visual Culture*, edited by Ann B. Shteir and Bernard V. Lightman. Dartmouth College Press, 2006.

Bicks, Caroline. *Cognition and Girlhood in Shakespeare's World*. Cambridge University Press, 2021.

Blackawton Primary School, S. Airzee, A. Allen, S. Baker, A. Berrow, C. Blair, M. Churchill, *et al.* "Blackawton Bees". *Biology Letters* 7.2 (2011): 168-72.

Blumenberg, Hans. *The Legitimacy of the Modern Age*. Translated by Robert M. Wallace. MIT Press, 1985.

Borberg, Gunnar. *The Man Who Organized Nature: The Life of Linnaeus*. Translated by Anna Paterson. Princeton University Press, 2023.

Bredekamp, Horst. "The Playfulness of Natural History". In Bredekamp, *The Lure of Antiquity and the Cult of the Machine*. Translated by Allison Brown. Markus Weiner, 1995.

Brown, James, Connor Plunkett and Agatha Yates. "Hooked: Robert Hooke's World of Intoxicants, 1672-83". https://www.intoxicatingspaces.org/maps/hooke/. Accessed 13 August 2024.

Caillois, Roger. *Man, Play and Games*. Translated by Meyer Barash. University of Illinois Press, 2001.

Campbell, Mary Baine. *Wonder and Science: Imagining Worlds in Early Modern Europe*. Cornell University Press, 1999.

Christianson, John Robert. *On Tycho's Island: Tycho Brahe and His Assistants*. Cambridge University Press, 2000.

Clucas, Stephen. "Poetic Atomism in Seventeenth-Century England: Henry More, Thomas Traherne and 'Scientific Imagination'". *Renaissance Studies* 5.3 (1991): 327-340.

Corneanu, Sorana. *Regimens of the Mind: Boyle, Locke, and the Early Modern Cultura Animi Tradition*. University of Chicago Press, 2011.

Covey, H. C. "A Return to Infancy: Old Age and the Second Childhood in History". *The International Journal of Aging and Human Development* 36.2 (1993): 81-90.

Cram, David, Jeffrey L. Forgeng, and Dorothy Johnston. "Introduction". In *Francis Willughby's Book of Games: A Seventeenth-Century Treatise on Sports, Games and Pastimes*, edited by David Cram, Jeffrey L. Forgeng, and Dorothy Johnson. Routledge, 2016.

Crane, Mary Thomas. *Losing Touch with Nature: Literature and the New Science in Sixteenth-Century England*. Johns Hopkins University Press, 2014.

Daston, Lorraine. "The Cold Light of Facts and the Facts of Cold Light: Luminescence and the Transformation of the Scientific Fact, 1600-1750". In *Signs of the Early Modern 2—17th Century and Beyond*, edited by David Lee Rubin. Rookwood Press, 1997.

Daston, Lorraine, and Peter Galison. *Objectivity*. Zone Books, 2007.

Daston, Lorraine, and Katherine Park. *Wonders and the Order of Nature, 1150-1750*. Zone Books, 1998.

Davies, Julie. "Botanizing at Badminton House: The Botanical Pursuits of Mary Somerset, First Duchess of Beaufort". In *Domesticity and the Making of Modern Science*, edited by Donald L. Opitz, Staffan Bergwik, and Brigitte Van Tiggelsen. Palgrave Macmillan, 2016.

Davis, Robert A. "Brilliance of a Fire: Innocence, Experience and the Theory of Childhood". *Journal of Philosophy of Education* 45.2 (20-11): 379-397.

Dear, Peter. "The Meanings of Experience". In *The Cambridge History of Science*, vol. 3. Edited by Katharine Park and Lorraine Daston. Cambridge University Press, 2006.

Dear, Peter. "A Philosophical Duchess: Understanding Margaret Cavendish and the Royal Society". In *Science, Literature, and Rhetoric in Early Modern England*, edited by Juliet Cummins and David Burchell. Routledge, 2007.

Darrigol, Olivier. *A History of Optics from Greek Antiquity to the Nineteenth Century*. Oxford University Press, 2012.

De Bary, Philip. *Thomas Reid and Scepticism: His Reliabilist Response*. Routledge, 2002.

DeFries, Brett. "Love, Capacity, and Traherne's Idea of the Book". *Studies in English Literature* 61.1 (2021): 103-26.

Delbourgo, James. *Collecting the World: Hans Sloane and the Origins of the British Museum*. Harvard University Press, 2017.

Devereaux, Johanna. "'Affecting the Shade': Attribution, Authorship, and Anonymity in 'An Essay in Defence of the Female Sex'". *Tulsa Studies in Women's Literature* 27.1 (2008): 17–37.

Dhaliwal, Ranjit. "Einstein's Tongue". *The Guardian*, Fri 14 Mar 2014. https://www.theguardian.com/artanddesign/picture/2014/mar/14/albert-einstein-tongue-photography. Accessed 04 January 2024.

DiMeo, Michelle. *Lady Ranelagh: The Incomparable Life of Robert Boyle's Sister*. Chicago University Press, 2020.

Dodd, Elizabeth S. "'Perfect Innocency in Creation' in the Writings of Thomas Traherne". *Literature and Theology* 29.2 (2015): 216-236.

Doddington, Christine, and Mary Hilton. *Child-Centred Education: Reviving the Creative Tradition*. Sage, 2007.

Doherty, Meghan C. "Discovering the 'True Form:' Hooke's *Micrographia* and the Visual Vocabulary of Engraved Portraits". *Notes and Records of the Royal Society* 66.3 (2012): 211-234.

Eco, Umberto. *The Aesthetics of Thomas Aquinas*. Translated by Hugh Bredlin. Harvard University Press, 1988.

Einstein, Albert. "Foreword". In Isaac Newton, *Opticks*. G Bell & Sons, 1931; reprinted by Dover Publications, 1952.

Espinasse, Margaret. *Robert Hooke*. University of California Press, 1956.

Evans, R. J. W., and Alexander Marr, eds., *Curiosity and Wonder from the Renaissance to the Enlightenment*. Ashgate, 2006.

Fara, Patricia. "Elizabeth Tollet: A New Newtonian Woman". *History of Science* 40 (2002): 169-187.

Fara, Patricia. *Pandora's Breeches: Women, Science and Power in the Enlightenment.* Pimlico, 2004.

Ferraro, Joanne M. "Childhood in Medieval and Early Modern Times". In *The Routledge History of Childhood in the Western World,* edited by Paula Fass. Routledge, 2013.

Findlen, Paula. "Between Carnival and Lent: The Scientific Revolution at the Margins of Culture". *Configurations* 6.2 (1998): 243-267.

Findlen, Paula. "Jokes of Nature and Jokes of Knowledge: The Playfulness of Scientific Discourse in Early Modern Europe". *Renaissance Quarterly* 43.2 (1990): 292-331.

Findlen, Paula. "Ludic Postscript". In *Ludi naturae: Spiele der Natur in Kunst und Wissenschaft,* edited by Natascha Adamowsky, Hartmut Böhme, and Robert Felfe. Wilhelm Fink, 2011.

Fudge, Erica. "Learning to Laugh". *Textual Practice* 17.2 (2003): 277-294.

Gadamer, Han Georg. *Truth and Method.* 2nd ed. Sheed and Ward, 1989.

Gal, Ofer, and Raz Chen-Morris. *Baroque Science.* Chicago University Press, 2013.

Gleick, James. *Isaac Newton.* Fourth Estate, 2003.

Golinski, J. V. "A Noble Spectacle: Phosphorus and the Public Cultures of Science in the Early Royal Society". *Isis* 80.1 (1989): 11-39.

Gómez, Pablo F. *The Experiential Caribbean: Creating Knowledge and Healing in the Early Modern Atlantic.* The University of North Carolina Press, 2017.

Gopnik, Alison, Andrew N. Meltzoff, and Patricia K. Kuhl. *The Scientist in the Crib: Minds, Brains and How Children Learn.* Harper Collins, 1999.

Gopnik, Alison. *The Philosophical Baby: What Children's Minds Tell Us About Truth, Love and the Meaning of Life.* Farrar, Strauss and Giroux, 2009.

Gorman, Cassandra. "Thomas Traherne and 'Feeling Inside the Atom'". In *Thomas Traherne and Seventeenth-Century Thought,* edited by Elizabeth S. Dodd and Cassandra Gorman. Boydell & Brewer, 2016.

Gowing, Laura. *Common Bodies: Women, Touch, and Power in Seventeenth-Century England.* Yale University Press, 2003.

Greengrass, Mark, Daisy Hildyard, Christopher D. Preston, and Paul J. Smith. "Science on the Move: Francis Willughby's Expeditions". In *Virtuoso by Nature: The Scientific Worlds of Francis Willughby FRS (1635-1672),* edited by Tim Birkhead. Brill, 2016.

Harkness, Deborah. *The Jewel House: Elizabethan London and the Scientific Revolution.* Yale University Press, 2008.

Harris, Anna. *A Sensory Education.* Routledge, 2020.

Harrison, Peter. "Curiosity, Forbidden Knowledge, and the Reformation of Natural Philosophy in Early Modern England". *Isis* 92 (2001): 265-290.

Harrison, Peter. *The Fall of Man and Foundations of Science*. Cambridge University Press, 2009.

Harrison, Timothy M. *Coming To: Consciousness and Natality in Early Modern England*. Chicago University Press, 2020.

Heilbron, John. "Domesticating Science in the Eighteenth Century". In *Science and the Visual Image in the Enlightenment*, edited by William R. Shea. Science History Publications, 2000.

Hein, Hilde. "Play as an Aesthetic Concept". *The Journal of Aesthetics and Art Criticism* 27.1 (1968): 67-71.

Henderson, Felicity. "Introduction" to "Unpublished Material from the Memorandum Book of Robert Hooke, Guildhall Library MS 1758"". *Notes and Records of the Royal Society* 61.2 (2007): 129-175.

Higonnet, Anne. *Pictures of Innocence: The History and Crisis of Ideal Childhood*. Thames and Hudson, 2008.

Hodacs, Hanna. "In the Field: Exploring Nature with Carolus Linnaeus". *Endeavour* 34.2 (2009): 45-9.

Horacek, Ivana. "Illuminating Methods, Picturing Instruments: Tycho Brahe's Instrumental Images". *Austrian History Yearbook* 52 (2021): 30-53.

Horkheimer, Max, and Theodor W. Adorno. *Dialectic of Enlightenment: Philosophical Fragments*, edited by Gunzelin Schmid Noerr, translated by Edmund Jephcott. Stanford University Press, 2007.

Howes, David. "The Misperception of the Environment: A Critical Evaluation of the Work of Tim Ingold". *Anthropological Theory* 22.4 (2002): 443-466.

Howes, David, and Constance Classen. *Ways of Sensing: Understanding the Senses in Society*. Routledge, 2014.

Huizinga, Johan. *Homo Ludens: A Study of the Play-Element in Culture*. Angelico Press, 2016.

Hünniger, Dominik. "Visible Labour? Productive Forces and Imaginaries of Participation in European Insect Studies, ca. 1680–1810". *Berichte zur Wissenschaftsgeschichte/History of Science and Humanities* 44.2 (2021): 180-210.

Hunter, Lynette, and Sarah Hutton, eds., *Women, Science and Medicine, 1500-1700: Mothers and Sisters of the Royal Society*. Sutton Publishing, 1997.

Hunter, Michael. "Alchemy, Magic and Moralism in the Thought of Robert Boyle". In *The British Journal for the History of Science* 23.4 (1990): 387-410.

Hunter, Michael. *Establishing the New Science: The Experience of the Royal Society*. Boydell & Brewer, 1995.

Hunter, Michael C. *Wicked Intelligence: Visual Art and the Science of Experiment in Restoration London*. Chicago University Press, 2013.

Hutton, Sarah. "Science and Natural Philosophy". In *The Routledge History of Women in Early Modern Europe*, edited by Amanda L. Capern. Routledge, 2019.

Hutton, Sarah. "Alchemy and Cultures of Knowledge among Early Modern Women". *Early Modern Women* 15.2 (2021): 93-102.

Jardine, Lisa. *The Curious Life of Robert Hooke: The Man who Measured London.* Harper Perennial, 2004a.

Jardine, Lisa. "The 2003 Wilkins Lecture: Dr Wilkins's Boy Wonders". *Notes and Records of the Royal Society of London* 58.1 (2004b): 107-29.

Jarvis, Erik J. "The Royal Society, Collective Vision, and Samuel Butler's 'The Elephant in the Moon'". In *Literature in the Age of Celestial Discovery: From Copernicus to Flamsteed*, edited by Judy A. Hayden. Palgrave Macmillan, 2016.

Jay, Martin. *Songs of Experience: Modern American and European Variations on a Universal Theme.* University of California Press, 2005.

Johnston, Carol Ann. "Heavenly Perspectives, Mirrors of Eternity: Thomas Traherne's Yearning Subject". *Criticism* 43.4 (2001): 377-405.

Kareem, Sarah Tindal. "Enlightenment Bubbles, Romantic Worlds". *The Eighteenth Century: Theory and Interpretation* 56.1 (2015): 85-104.

Kaseman, Melissa. *Preschool Pocket Treasures.* http://www.melissakaseman.com/preschool-pocket-treasures. Accessed 08 October 2022.

Kelly, J. N. D. *Early Christian Doctrines.* London, 1958.

Kidd, Stephen E. *Play and Aesthetics in Ancient Greece.* Cambridge University Press, 2019.

Kirch, Susan A., and Michele Amoroso. *Being and Becoming Scientists Today: Reconstructing Assumptions about Science and Science Education to Reclaim a Learner–Scientist Perspective.* Brill, 2016.

Klein, Jürgen. "Francis Bacon's *Scientia Operativa*, The Tradition Of The Workshops, And The Secrets Of Nature". In *Philosophies of Technology: Francis Bacon and his Contemporaries*, vol. 1. Edited by Claus Zittel, Romano Nanni, Gisela Engel, and Nicole Karafyllis. Brill, 2008.

Kohan, Walter O. "Childhood, Philosophy, and the Polis: Exclusion and Resistance". In *Philosophy of Childhood Today: Exploring the Boundaries*, edited by David Kennedy and Brock Bahler. Lexington Books, 2017.

Kuhn, Deanna. "Piaget's Child as Scientist". In *Piaget's Theory: Prospects and Possibilities*, edited by Harry Beilin and Peter B. Pufall. Lawrence Erlbaum Associates, 1992.

Lamb, Edel. *Reading Children in Early Modern Culture.* Palgrave Macmillan, 2018.

Larson, Katherine R. "'Certein childeplayes remembred by the fayre ladies': Girls and Their Games". In *Gender and Early Modern Constructions of Childhood*, edited by Naomi Miller and Naomi Yavneh. Ashgate, 2011.

Laszlo, Pierre. "Macroscope: Science as Play". *American Scientist* 92.5 (2004): 398-400.

Laudan, L. L. "Thomas Reid and the Newtonian Turn of British Methodological Thought". In *The Methodological Heritage of Newton*, edited by Robert E. Butts and John W. Davis. University of Toronto Press, 1970.

Lawson, Ian. "Crafting the Microworld: How Robert Hooke Constructed Knowledge about Small Things". Notes and Records of the Royal Society 70.1 (2016): 23-44.

Lekies, Kristi S., and Thomas H. Beery. "Everyone Needs a Rock: Collecting Items from Nature in Childhood". *Children, Youth and Environments* 23.3 (2013): 66–88.

Leong, Elaine. *Recipes and Everyday Knowledge: Medicine, Science, and the Household in Early Modern England*. Chicago University Press, 2018.

Lewis, Rhodri. "Of 'Origenian Platonisme': Joseph Glanvill on the Pre-existence of Souls". *Huntington Library Quarterly* 69.2 (2006): 267-300.

Long, Kathleen P., ed. *Gender and Scientific Discourse in Early Modern Culture*. Ashgate, 2010.

Lotto, Beau. "Why Science is Like Play". *CNN*, 11 November 2012. https://edition.cnn.com/2012/11/11/opinion/lotto-ted-science-play/index.html. Accessed 14 August 2024.

Lund, Roger G. "'More strange than true': Sir Hans Sloane, King's *Transactioneer*, and the Deformation of English Prose". *Studies in Eighteenth-Century Culture* 14 (1985): 213-30.

Lyons, Tony. "Play and Toys in the Educational Work of Richard Lovell Edgeworth 1744–1817". *Irish Educational Studies* 20.1 (2001): 310–20.

Matthews, Gareth B. *The Philosophy of Childhood*. Harvard University Press, 1996.

Matthews, Gareth B. *Philosophy and the Young Child*. Harvard University Press, 1980.

Merchant, Carolyn. *The Death of Nature: Women, Ecology and the Scientific Revolution*. Harper and Row, 1980.

Moshenska, Joe. *Iconoclasm as Child's Play*. Stanford University Press, 2019.

Moyer, Ann. "The Astronomers' Game: Astrology and University Culture in the Fifteenth and Sixteenth Centuries". *Early Science and Medicine* 4. 3 (1999): 228-50.

Mûelenaere, Gwendoline de "The Art of Learning: Illustrated Lecture Notebooks at the Old University of Louvain". In *Scientific Visual Representations in History*, edited by Matteo Valleriani, Giulia Giannini, and Enrico Giannetto. Springer, 2023.

Murphy, Kathryn. "No Things but in Thoughts: Traherne's Poetic Realism". In *Thomas Traherne and Seventeenth-Century Thought*, edited by Elizabeth S. Dodd and Cassandra Gorman. Boydell & Brewer, 2016.

Myers, Mitzi. "'Anecdotes from the Nursery' in Maria Edgeworth's *Practical Education* (1798): Learning from Children 'Abroad and At Home'". *The Princeton University Library Chronicle* 60.2 (1999): 220-250.

Nelson, Holly Faith, and Sharon Alker. "Virtual Reality, Role Play and World Building in Margaret Cavendish's Literary War Games". In *Games and War in Early Modern English Literature: From Shakespeare to Swift*, edited by Holly Faith Nelson and James William Daems. Amsterdam University Press, 2019.

Nicolson, Marjorie Hope. *The Breaking of the Circle: Studies in the Effect of the 'New Science' upon Seventeenth-Century Poetry*. Revised ed. Columbia University Press, 1960.

Oldroyd, D. R."Some 'Philosophicall Scribbles' attributed to Robert Hooke". *Notes and Records of the Royal Society of London* 35.1 (1980): 17-32.

Olsen, Glending. "Play as Play: A Medieval Ethical Theory of Performance and the Intellectual Context of the Tretise of Miraclis Pleyinge". *Viator* 26 (1995): 195-222.

Opitz, Donald L., and Staffan Bergwik and Brigitte Van Tiggelsen. "Introduction: Domesticity and the Historiography of Science". In *Domesticity and the Making of Modern Science*. Palgrave Macmillan, 2016.

Orme, Nicholas. *Medieval Children*. Yale University Press, 2001.

Parrish, Susan Scott. *American Curiosity: Cultures of Natural History in the Colonial British Atlantic World*. The University of North Carolina Press, 2006.

Partner, Jane. *Poetry and Vision in Early Modern England*. Palgrave Macmillan, 2018.

Pesic, Peter. "Wrestling with Proteus: Francis Bacon and the 'Torture' of Nature". *Isis* 90.1 (1999): 81-94.

Piaget, Jean. *The Language and Thought of the Child*. Routledge, 2001.

Picciotto, Joanna. *Labours of Innocence in Early Modern England*. Harvard University Press, 2010.

Pitkin, Barbara. "'The Heritage of the Lord': Children in the Theology of John Calvin". In The *Child in Christian Thought*, edited by Marcia J. Bunge. William B. Eerdmans Publishing, 2000.

Pollock, Linda. *Forgotten Children: Parent-child Relations from 1500 to 1900*. Cambridge University Press, 1983.

Poole, Kirsten. "Bacon and Allegory". In *The Palgrave Handbook of Early Modern Literature and Science*, edited by Howard Marchitello and Evelyn Tribble. Palgrave Macmillan, 2017.

Quehen, Hugh. "Butler, Samuel (bap. 1613, d. 1680), poet". *Oxford Dictionary of National Biography*.https://www.oxforddnb.com/view/10.1093/. Accessed 05 September 2024.

Raymo, Chester. "Science as Play". *Science Education* 57.3 (1973): 279-89.

Reitsma, Ella. *Maria Sibylla Merian and Daughters: Women of Art and Science*. Waanders Publishers, 2008.

Rendall, Jane. "'Elementary Principles of Education': Elizabeth Hamilton, Maria Edgeworth and the Uses of Common Sense Philosophy". *History of European Ideas* 39.5 (2012): 613-30.

Richardson, Harley. "How we learned to teach 'small children'". *History of Education Blog*. https://historyofeducation.net/2022/01/13/how-we-learned-to-teach-small-children/. Accessed 17 June 2024.

Rimmer, Chad Michael. *Greening the Children of God: Thomas Traherne and Nature's Role in the Moral Formation of Children*. The Lutterworth Press, 2021.

Robinson, Ken. "The Skepticism of Butler's Satire on Science: Optimistic or Pessimistic?" *Restoration: Studies in English Literary Culture, 1660–1700* 7.1 (1983): 1-7.

Rowland, Ann Weirda. *Romanticism and Childhood: The Infantilization of British Literary Culture*. Cambridge University Press, 2012.

Rutter, Carol Chillington. *Shakespeare and Child's Play: Performing Lost Boys on Stage and Screen*. Routledge, 2007.

Sabra, A. I. *Theories of Light from Descartes to Newton*. 2nd ed. Cambridge University Press, 1981.

Sarasohn, Lisa T. *The Natural Philosophy of Margaret Cavendish: Reason and Fancy During the Scientific Revolution*. Johns Hopkins University Press, 2010.

Sawday, Jonathan. *The Body Emblazoned: Dissection and the Human Body in Renaissance Culture*. Routledge, 1995.

Schaffer, Simon. "Glass Works: Newton's Prisms and the Uses of Experiment". In *The Uses of Experiment: Studies in the Natural Sciences*, edited by David Gooding, Trevor Pinch, and Simon Schaffer. Cambridge University Press, 1985.

Schaffer, Simon. "A Science whose Business is Bursting: Soap Bubbles as Commodities in Classical Physics". In *Things That Talk: Object Lessons from Art and Science*, edited by Lorraine Daston. Zone Books, 2004.

Schiebinger, Londa. *Plants and Empire: Colonial Bioprospecting in the Atlantic World*. Harvard University Press, 2004.

Scott, Dominic. *Recollection and Experience: Plato's Theory of Learning and Its Successors*. Cambridge University Press, 1995.

Secord, Jim A. "Newton in the Nursery: Tom Telescope and the Philosophy of Tops and Balls, 1761–1838". *History of Science* 23.2 (1985): 127-151.

Sekimoto, Sachi and Christopher Brown. "Introduction: Feeling Race". In *Race and the Senses: The Felt Politics of Racial Embodiment*, edited by Sachi Sekimoto and Christopher Brown. Routledge, 2020.

Shahar, Shulamith. *Childhood in the Middle Ages*. Routledge, 1990.

Shaheen, Jonathan L. "A Vitalist Shoal in the Mechanist Tide: Art, Nature, and 17th-Century Science". *Philosophies* 7.5 (2022): 111-31.

Shapin, Steven. "The House of Experiment in Seventeenth-Century England". *Isis* 79.3 (1988): 373-404.

Shapin, Steven. "The Invisible Technician". *American Scientist* 77.6 (1989): 554-563.

Shapin, Steven, and Simon Schaffer. *Leviathan and the Air-Pump: Hobbes, Boyle, and the Experimental Life*. Princeton University Press, 1985.

Shapin, Steven. "The Philosopher and the Chicken: On the Dietetics of Disembodied Knowledge". In *Science Incarnate: Historical Embodiments of Natural Knowledge*, edited by Christopher Lawrence and Steven Shapin. University of Chicago Press, 1998.

Shapin, Steven. *A Social History of Truth: Civility and Science in Seventeenth-Century England*. University of Chicago Press, 1994.

Simon, David Carroll. *Light without Heat: The Observational Mood from Bacon to Milton*. Cornell University Press, 2018.

Smith, Hilda L. "Margaret Cavendish and the Microscope as Play". In *Men, Women and the Birthing of Modern Science*, edited by Judith P. Zinsser. Northern Illinois University Press, 2005.

Smith, Kurt. "Hyperaspistes". In *The Cambridge Descartes Lexicon*, edited by Lawrence Nolan. Cambridge University Press, 2015.

Smith, Pamela H. *The Body of the Artisan: Art and Experience in the Scientific Revolution*. University of Chicago Press, 2004.

Spiller, Elizabeth. *Science, Reading, and Renaissance Literature: The Art of Making Knowledge*. Cambridge University Press, 2004.

Stafford, Barbara. *Artful Science: Enlightenment Education and the Eclipse of Visual Education*. The MIT Press, 1994.

Stark, Ryan J. "Margaret Cavendish and Composition Style". *Rhetoric Review* 17.2 (1999): 264-281.

Steenberg, M.C. "Children in Paradise: Adam and Eve as 'Infants' in Irenaeus of Lyons". *Journal of Early Christian Studies* 12.1 (2004): 1-22.

Stone, Lawrence. *The Family, Sex and Marriage in England, 1500-1800*. Harper and Row, 1977.

Stone, Denise L. "Children's Collections and the Art Museum", *Visual Arts Research* 34.1 (2008): 75–86.

Stortz, Martha Ellen. "'Where or When Was Your Child Innocent?': Augustine on Childhood". In *The Child in Christian Thought*, edited by Marcia J. Bunge. William B. Eerdmans Publishing, 2000.

Sullivan, Erin. "The Passions of Thomas Wright: Renaissance Emotion Across Body and Soul". In *The Renaissance of Emotion: Understanding Affect in Shakespeare and his Contemporaries*, edited by Richard Meek, and Erin Sullivan. Manchester University Press, 2015.

Swann, Elizabeth L. "From Philosopher's Stone to Phosphorus: Robert Boyle's Illuminating Experiments". In *The Poesy of Scientia in Early Modern England*, edited by Subha Mukherji and Elizabeth L. Swann. Palgrave Macmillan, 2024.

Thomas, Keith. "Age and Authority in Early Modern England". *Proceedings of the British Academy* 62 (1976): 205-48.

Thomas, Keith. "Children in Early Modern England". In *Children and Their Books*, edited by Gillian Avery and Julia Briggs. Oxford University Press, 1989.

Tribble, Evelyn. "Watery Knowledge on the Early Modern Stage". *Shakespeare Studies* 49 (2021): 119-127.

Wall, John. "Childhood in Western Philosophy". In *The SAGE Encyclopedia of Children and Childhood Studies*, edited by Daniel Thomas Cook. SAGE Publications, 2020.

Wall, Wendy. *Recipes for Thought: Knowledge and Taste in the Early Modern English Kitchen*. University of Pennsylvania Press, 2015.

Walters, Lisa. "The Philosophy and Literature of Childhood Cognition: John Milton and Margaret Cavendish". In *Literary Cultures and Medieval and Early*

Modern Childhoods, edited by Naomi J. Miller and Diane Purkiss. Palgrave Macmillan, 2019.

Watson, Robert. *Back to Nature: The Green and the Real in the Late Renaissance.* University of Pennsylvania Press, 2006.

Weeks, Sophie. "The Role of Mechanics in Francis Bacon's Great Instauration". In *Philosophies of Technology: Francis Bacon and his Contemporaries*, vol. 1. Edited by Claus Zittel, Romano Nanni, Gisela Engel, and Nicole Karafyllis. Brill, 2008.

Whitbread, Nanette. *The Evolution of the Nursery-Infant School: A History of Infant Education in Britain, 1800-1970.* Routledge, 1972.

Whitmer, Kelly. *The Halle Orphanage as Scientific Community: Observation, Eclecticism, and Pietism in the Early Enlightenment.* The University of Chicago Press, 2015.

Whitmer, Kelly. "Reimagining the 'Nature of Children': Realia, Reform, and the Turn to Pedagogical Realism in Central Europe c. 1600–1700". *The Journal of the History of Childhood and Youth* 12.1 (2019): 113-135.

Wilding, Nick, "Galileian Angels". In *Conversations with Angels: Essays Towards a History of Spiritual Communication, 1100–1700*, edited by Joad Raymond. Palgrave Macmillan, 2011.

Wilkin, Rebecca. "Descartes, Individualism, and the Fetal Subject". *Differences* 19.1 (2008): 96-127.

Willmott, Richard. *The Voluble Soul: Thomas Traherne's Poetic Style and Thought.* The Lutterworth Press, 2021.

Wilson, Catherine. *The Invisible World: Early Modern Philosophy and the Invention of the Microscope.* Princeton University Press, 1997.

Witmore, Michael. *Culture of Accidents: Unexpected Knowledges in Early Modern England.* Stanford University Press, 2002.

Witmore, Michael. *Pretty Creatures: Children and Fiction in the English Renaissance.* Cornell University Press, 2007.

Wood, Paul. "Thomas Reid and the Culture of Science". In *The Cambridge Companion to Thomas Reid*, edited by Terence Cuneo and Renè van Woudenberg. Cambridge University Press, 2004.

Woodland, Patrick. "Beale, John". *Oxford Dictionary of National Biography.* https://doi-org.ezphost.dur.ac.uk/10.1093/ref:odnb/1802.

Wragge-Morley, Alexander. *Aesthetic Science: Representing Nature in the Royal Society of London, 1650-1720.* Chicago University Press, 2020.

Zagorin, Perez. "Francis Bacon's Concept of Objectivity and the Idols of the Mind". *BJHS* 34 (2001): 379-93.

Zinsser, Judith, ed. *Men, Women, and the Birthing of Modern Science.* Cornell University Press, 2005.

Index[1]

A

Accident, *see* Chance
Adorno, Theodor W., 11, 117
Adulthood, *see* Maturity
Aesthetics, aesthetic judgement, 53, 55, 55n53, 57
Affect, *see* Passions, and affects
Agrippa, Cornelius, 46, 46n17
Allen, Barry, 65–67
Anatomy, 102, 103
Antiquarianism, 60n69
Apprentices, *see* Assistants, scientific
Aquinas, Thomas, 52, 52n45, 53
Ariès, Philippe, 16, 16n51
Aristotle, 22, 25, 26, 26n18, 35, 51–53, 58n63, 66, 76
Artisans, 11, 44, 49, 86
Ascham, Roger, 25, 37
Assistants, scientific, 17, 72, 73, 84, 86–88, 104, 105
Astronomy, 6, 76n50, 86n4, 87, 88

Augustine, St., 5, 5n14, 5n15, 10n33, 42, 43
   childrens lack of, 43
   in relation to play, 5

B

Bacon, Francis, 4–8, 4n13, 7n21, 10, 10n36, 11, 13, 15, 26, 30, 32, 36, 40, 48, 48n24, 49, 53, 55n55, 56, 56n59, 59, 66, 68, 68n22, 77, 77n53, 77n54, 78, 91n31, 94–97, 94n42, 96n46, 97n48, 100, 110, 117, 117n27, 117n28, 119
Batholomaeus, Angelicus, 16–17
Beale, John, 26, 27, 27n23, 30, 37, 84, 89, 90
Bible, the
   Adam and Eve, 40, 69
   Book of Job, 5

---

[1] Note: Page numbers followed by 'n' refer to notes.